ELEMENTS OF WAVE MECHANICS

ELEMENTS OF
WAVE MECHANICS

BY

N. F. MOTT
M.A., D.Sc., F.R.S.
Cavendish Professor of Experimental Physics
University of Cambridge

CAMBRIDGE
AT THE UNIVERSITY PRESS
1962

PUBLISHED BY
THE SYNDICS OF THE CAMBRIDGE UNIVERSITY PRESS

Bentley House, 200 Euston Road, London, N.W. 1
American Branch: 32 East 57th Street, New York 22, N.Y.
West African Office: P.O. Box 33, Ibadan, Nigeria

First Printed 1952
Reprinted 1958
1962

First printed in Great Britain by John Wright & Sons Ltd., Bristol
Reprinted by offset-litho by John Dickens and Conner, Ltd., Northampton

PREFACE

This book is written to replace the author's *An Outline of Wave Mechanics*, published by the Cambridge University Press in 1930 and now out of print. It is intended for students in the final year of an honours course of experimental physics, and also as an introduction to more advanced text-books for those who intend to specialise in the subject. With such readers in view, and in order to keep the book reasonably short, I have not hesitated to omit some mathematical developments that can easily be found in other text-books. The book attempts to build up the elementary theory from experimental facts, and to show how simple problems can be solved. A certain number of examples are included.

N. F. MOTT

BRISTOL
January, 1951

13306

PREFACE

P. J. TAYLOR

Auckland

CONTENTS

CHAPTER I

THE DIFFERENTIAL EQUATIONS OF WAVE MECHANICS

1. DIFFERENTIAL EQUATIONS OF THE SECOND ORDER

Many problems in wave mechanics can be reduced to the solution of a differential equation of the type

$$\frac{d^2y}{dx^2}+f(x)\,y = 0, \tag{1}$$

and a thorough understanding of this equation is essential to the student of the subject; $f(x)$ is a known function of x, and by means of the equation it is possible to plot y against x when the values of y and of dy/dx are given for one arbitrary value of x. An equivalent statement is that two independent solutions, y_1 and y_2, exist and that $Ay_1 + By_2$ is the general solution. Methods[†] exist of obtaining these solutions graphically, which can be utilised when $f(x)$ is given as a plotted curve or tabulated.

The simplest case of equation (1) is that in which $f(x)$ is a constant; if $f(x)$ is a positive constant, we may write

$$f(x) = k^2,$$

and two independent solutions then exist, $\cos kx$ and $\sin kx$. The general solution is

$$y = A \cos kx + B \sin kx$$

or
$$y = a \cos (kx + \epsilon),$$

where A, B, a and ϵ are arbitrary constants. The solution is then oscillating (Fig. 1a).

[†] Cf., for example, D. R. Hartree, *Proc. Manchr. lit. phil. Soc.* LXXVII, 91, 1933; W. F. Manning and J. Millman, *Phys. Rev.* LIII, 673, 1938; M. V. Wilkes, *Proc. Camb. phil. Soc.* XXXVI, 204, 1940; L. Fox and E. T. Goodwin, *Proc. Camb. phil. Soc.* XLV, 373, 1949.

If $f(x)$ is constant and negative, we set

$$f(x) = -\gamma^2$$

and the solutions are then $e^{-\gamma x}$ and $e^{\gamma x}$, with the general solution

$$y = Ae^{\gamma x} + Be^{-\gamma x}.$$

These solutions also are illustrated in Fig. 1b.

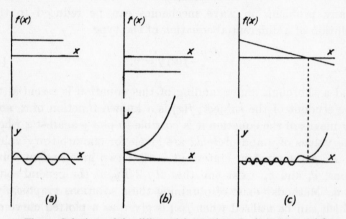

Fig. 1. Solutions of the differential equation $y'' + f(x)y = 0$. (a) for $f(x) = k^2$, (b) for $f(x) = -\gamma^2$, (c) for an arbitrary form of $f(x)$ which changes sign, the solution for negative values of x taking either of the forms shown in (b).

In the general case where $f(x)$ is not a constant, it is easy to show that, if $f(x)$ is positive, y is an oscillating function, while if $f(x)$ is negative, y is of exponential form. For if $f(x)$ is positive, y and d^2y/dx^2 have the opposite sign. Therefore, if we consider any point A on the curve for which y is positive, the curve bends in the direction shown in Fig. 2a, getting steeper and steeper until it crosses the axis when it will begin to bend in the opposite direction. If, on the other hand, $f(x)$ is negative, y and d^2y/dx^2 are of the same sign, and the slope at a point such as A will increase, giving an exponentially increasing curve as shown in Fig. 2b.

The general form of the solution y for a function $f(x)$ which changes sign is as shown in Fig. 1c. When $f(x)$ becomes negative y goes over to the exponential form. It should be emphasised that there will always be one solution which decreases exponentially, but that the general solution will increase. If one stipulates that the solution should be that which decreases exponentially, this determines the phase of the oscillations in the range of x for which oscillations occur.

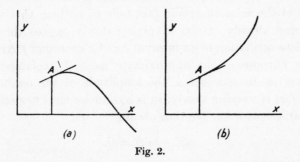

Fig. 2.

A useful method exists of determining approximate solutions of the differential equation (1), for the case where $f(x)$ does not vary too rapidly with x. This is known in the literature as the Wentzel-Kramers-Brillouin† (W.K.B.) method, though apparently first given by Jeffreys.‡ In the discussion given here we shall confine ourselves to the case where $f(x)$ is positive.

In order to obtain this approximate solution we set

$$y = \alpha e^{i\beta}. \tag{2}$$

Here α and β are both functions of x; α represents the amplitude of the oscillations, β the phase. On substituting into (1) we obtain

$$\alpha'' + i(\alpha\beta'' + 2\alpha'\beta') + (f - \beta'^2)\alpha = 0,$$

† G. Wentzel, Z. Phys. xxxviii, 518–29, 1926; H. A. Kramers, Z. Phys. xxxix, 828–40, 1926; L. Brillouin, C.R. Acad. Sci., Paris, clxxxiii, 24–6, 1926.

‡ H. Jeffreys, Proc. Lond. math. Soc. (2), xxiii, 428–36, 1925; Phil. Mag. xxxiii, 451–6, 1942.

where a dash denotes differentiation with respect to x. We are at liberty to impose one arbitrary condition on the two functions α, β; we set therefore

$$f - \beta'^2 = 0,$$

so that

$$\beta = \int_{x_0}^{x} f^{\frac{1}{2}} dx, \tag{3}$$

where x_0 is an arbitrary constant. This gives us correctly the phase of the solution and in fact tells us nothing that we did not know already; if $f(x)$ is varying slowly, y goes through a complete oscillation in an interval Δx of x such that $f^{\frac{1}{2}}\Delta x = 2\pi$.

The purpose of the approximate method explained here, however, is to estimate α, the amplitude of the oscillations. Since $f(x)$ is varying slowly, so is α, and we thus neglect α'' in comparison with α'. We thus obtain

$$\alpha\beta'' + 2\alpha'\beta' = 0,$$

whence on integrating

$$\ln \alpha + \tfrac{1}{2}\ln \beta' = \text{const.}$$

This gives

$$\alpha = \text{const.}\,(\beta')^{-\frac{1}{2}},$$

or, substituting for β' from (3),

$$\alpha = \text{const.}\,f^{-\frac{1}{4}}.$$

The approximate solution (2) is thus

$$y = \text{const.}\,f^{-\frac{1}{4}}\exp\left\{i\int_{x_0}^{x} f^{\frac{1}{2}} dx\right\}.$$

It follows that the amplitude of the oscillations *increases* as f becomes smaller and the wavelength increases. This is shown in Fig. 1c.

Exercise

Show that if $f = x^{-4}$, the above solution is exact.

In certain cases it is possible to obtain a solution of (1), or of the more general equation

$$\frac{d^2y}{dx^2} + g(x)\frac{dy}{dx} + f(x)y = 0,$$

in the form of a power series. As an example of the method we give the solution of Bessel's equation

$$\frac{d^2y}{dx^2} + \frac{1}{x}\frac{dy}{dx} + \left(1 - \frac{n^2}{x^2}\right)y = 0. \tag{4}$$

We set for y
$$y = x^\rho(a_0 + a_1 x + a_2 x^2 + \ldots). \tag{5}$$

When this series is inserted in the equation, the coefficient of every power of x must vanish. The lowest power of x which occurs is that of $\rho - 2$; the coefficient of $x^{\rho-2}$ must vanish. This gives the quadratic equation known as the indicial equation

$$\rho^2 - n^2 = 0$$

or
$$\rho = \pm n.$$

There are thus two possible values of ρ; these give the two independent solutions of the equation.

The coefficient of $x^{\rho+s}$ is

$$a_{s+2}\{(\rho + s + 2)^2 - n^2\} + a_s.$$

Since this must vanish we obtain a relation between a_{s+2} and a_s. Thus from a_0 we obtain a_2; from a_2 we obtain a_4; and so on. The student will easily see that the odd coefficients a_1, a_3, etc., vanish.

It is of interest to discuss the form of the solutions of (4) for large n by means of the W.K.B. method. The equation (4) may be reduced to the standard form (1) by means of the substitution

$$y = x^{-\frac{1}{2}}z,$$

giving for z
$$\frac{d^2z}{dx^2} + \left(1 - \frac{n^2 - \frac{1}{4}}{x^2}\right)z = 0.$$

It will be seen that for $x > \sqrt{(n^2 - \frac{1}{4})}$ the solutions are oscillating, and for large x behave like $\cos x$ and $\sin x$ or $e^{\pm ix}$. For $x < \sqrt{(n^2 - \frac{1}{4})}$, on the other hand, they are exponential. The solution in series for which the first term in (5) is x^n clearly corresponds to the solution which decreases as x decreases, while that which begins with x^{-n} corresponds to the increasing solution. The two solutions are illustrated in Fig. 3.

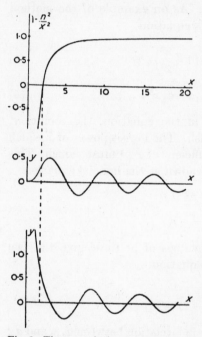

Exercise

Discuss the form of the solution for small positive values of x of

$$z'' + \left(1 + \frac{n^2}{x^2}\right)z = 0,$$

both by means of the indicial equation, and by the W.K.B. method. Show that the number of oscillations between $x = 0$ and any finite value of x is infinite. Why is this not the case for the equation

$$z'' + \left(1 + \frac{n}{x}\right)z = 0\,?$$

Fig. 3. The two solutions of (4) for $n = 2$.

2. WAVE EQUATIONS

In most forms of wave motion in one dimension, the displacement Ψ of the vibrating medium at a point x in space satisfies an equation of the form

$$\frac{\partial^2 \Psi}{\partial t^2} = v^2 \frac{\partial^2 \Psi}{\partial x^2}. \tag{6}$$

v (the wave velocity) is a constant which depends on the medium. Examples are:

(a) *The vibrations of a string* of tension T and mass ρ per unit length. Ψ is here the lateral displacement, and x the distance measured along the string. The proof of this relation, together with the formula for v,

$$v^2 = T/\rho,$$

can be found in any text-book on vibrations, e.g. Coulson's *Waves*, Chapter II.

(b) *Sound waves.* Here Ψ is the condensation, or $\Delta\rho/\rho$, where ρ is the density.

(c) *Electromagnetic waves.* In a plane polarised wave, the state of the medium is described by the electric vector E and the magnetic vector H, which are perpendicular to each other and to the direction of propagation. When the latter is along the x-axis they satisfy the equations

$$c\frac{\partial E}{\partial x} = \frac{\partial H}{\partial t}, \quad \frac{c}{\kappa}\frac{\partial H}{\partial x} = \frac{\partial E}{\partial t}, \tag{7}$$

where c is the velocity of light in a vacuum and κ the dielectric constant. Eliminating H between these equations we obtain

$$\frac{c^2}{\kappa}\frac{\partial^2 E}{\partial x^2} = \frac{\partial^2 E}{\partial t^2}. \tag{8}$$

The general solution of equation (6), as may easily be verified, is

$$\Psi = F(x - vt) + G(x + vt),$$

where F and G are any arbitrary functions whatever. This solution means that some quite arbitrary form of displacement, $F(x)$, is moving with velocity v to the right without change of form, and another function $G(x)$ moves to the left. This is illustrated in Fig. 4.

Particular solutions are:

(i) $$\Psi = A \sin k(x - vt),$$

which represents a simple harmonic wave of wavelength $\lambda = 2\pi/k$;

(ii) $\qquad \Psi = A\{\sin k(x-vt) + \sin k(x+vt)\}$

$\qquad\qquad = 2A \sin kx \cos kvt,$

which represents a standing wave.

The generalisation to three dimensions of equation (6) is

$$\frac{\partial^2 \Psi}{\partial t^2} = v^2 \nabla^2 \Psi, \qquad (9)$$

where ∇^2 is defined as

$$\nabla^2 = \frac{\partial^2}{\partial x^2} + \frac{\partial^2}{\partial y^2} + \frac{\partial^2}{\partial z^2}.$$

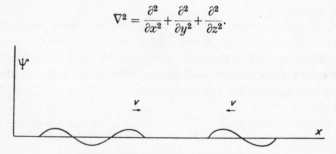

Fig. 4. General solution of equation (6).

The solution representing a plane wave moving in the direction defined by the direction-cosines (l, m, n) is

$$\Psi = A \sin \{k(lx + my + nz - vt)\}.$$

It is easily verified that this is a solution of (9), making use of the relation $l^2 + m^2 + n^2 = 1$.

The wave velocity v may vary from point to point in space. Thus in the case of the vibrating string the density ρ may vary along the length of the string; in the case of electromagnetic radiation, the light may pass from a medium of one refractive index to another. Two cases of interest present themselves, (i) that of a sharp boundary and (ii) that of a gradual change.

(i) There will be a sharp change in v, for example at the boundary between glass and air, or at a point where two strings of different density are joined together. Any wave incident normally on such a boundary will be partially reflected.

We shall now calculate the amplitude of the reflected wave at a boundary where the wave number changes suddenly from k to k'. It is convenient to use the complex exponential form for the function representing the waves, it being understood that the wave amplitude is actually represented by the real (or imaginary) part of the function written down.† We then write for the incident wave‡

$$e^{i(kx-\omega t)},$$

for the reflected wave

$$Ae^{i(-kx-\omega t)},$$

and for the transmitted wave

$$Be^{i(k'x-\omega t)}.$$

The problem is to calculate A and B. $\omega/2\pi$ is here the frequency, which, of course, cannot change when the wave goes from one medium into another.

It is convenient to choose the plane $x = 0$ for the boundary at which v and hence $k(= \omega/v)$ changes. We then have for the wave amplitude

$$\Psi = (e^{ikx} + Ae^{-ikx})\,e^{-i\omega t} \quad x < 0,$$
$$= Be^{ik'x}e^{-i\omega t} \qquad\qquad x > 0.$$

The values of A and B are determined by the boundary conditions applicable at $x = 0$. For the case of the string, it is obvious that these are that Ψ and $\partial\Psi/\partial x$ must be continuous; in other words, there is no kink in the string. In the case where Ψ is the electric vector E of a plane polarised wave falling normally on a reflecting surface, it may be shown§ that the same conditions are satisfied. The condition that Ψ is continuous then gives

$$1 + A = B,$$

and that $\partial\Psi/\partial x$ is continuous,

$$k(1 - A) = k'B.$$

---

† As will be shown in Chap. II, in wave mechanics we actually represent the amplitude by a complex function.

‡ Alternatively we could use the complex conjugate; in this chapter we shall use the convention that the time factor is $e^{-i\omega t}$.

§ Cf., for example, J. A. Stratton, *Electromagnetic Theory*, New York, 1941, p. 35.

On solving for A and B we find

$$A = (k - k')/(k + k'), \quad B = 2k/(k + k').$$

If we define by R the proportion of the incident energy reflected, then R can be equated to A^2, so that

$$R = (k - k')^2/(k + k')^2. \tag{10}$$

Since $k'/k = \mu$, where μ is by definition the refractive index, this may be written

$$R = (1 - \mu)^2/(1 + \mu)^2.$$

For glass, for example ($\mu \sim 1 \cdot 5$), this gives $R \sim 0 \cdot 04$.

Exercise

Verify, for waves in strings and for electromagnetic waves, that the flow of energy in the reflected and transmitted waves is equal to that in the incident wave.

A particularly interesting case arises when k', the wave number in the medium on which the wave is falling, is imaginary. This can arise, for instance, for electromagnetic waves in a medium containing free electrons; if N is the number of such electrons per unit volume, the refractive index μ is given by†

$$\mu^2 = 1 - \frac{Ne^2}{\pi m v^2},$$

where e is the electronic charge; for low enough values of the frequency ν the right-hand side becomes negative. For such cases we may write

$$k' = i\gamma,$$

where γ is real and positive. For the transmitted wave ($x > 0$) we have then two alternative forms

$$e^{-\gamma x} e^{-i\omega t}, \quad e^{\gamma x} e^{-i\omega t}.$$

Obviously the latter is not admissible, since it *increases* indefinitely as x increases. We therefore take for $x > 0$

$$\Psi = B e^{-\gamma x} e^{-i\omega t}.$$

† This equation is discussed further in Chap. IV, § 10.

The boundary conditions now give

$$1 + A = B,$$

$$ik(1 - A) = -\gamma B,$$

whence
$$A = \frac{k - i\gamma}{k + i\gamma}.$$

The modulus of A is unity; in other words, A is of the form $e^{i\alpha}$; thus the amplitude of the reflected wave is now unity. Therefore, when a wave is incident on a medium of imaginary refractive index, it is totally reflected.

Exercise

Investigate the reflection of waves from a slab of material in which the refractive index is imaginary. Take for the boundaries of the slab $x = 0$ and $x = a$, and set

$$\Psi = (e^{ikx} + Ae^{-ikx}) e^{-i\omega t} \quad x < 0,$$

$$= (Be^{\gamma x} + Ce^{-\gamma x}) e^{-i\omega t} \quad 0 < x < a,$$

$$= De^{ikx} e^{-i\omega t} \quad a < x.$$

Show that if $e^{-\gamma a} \ll 1$, then approximately

$$|D| = \frac{4k\gamma}{k^2 + \gamma^2} e^{-\gamma a},$$

and thus that the intensity penetrating the slab decreases as $e^{-2\gamma a}$ as a is increased.

(ii) If there is no sharp change in k, but a gradual change from point to point, then no partial reflection occurs; but a beam of waves will be bent as it traverses the medium. A formula for the radius of curvature of a beam of waves in traversing a medium will be important. Fig. 5 shows such a beam. ABC, $A'B'C$ are wave fronts one wavelength apart. We require to know the radius of curvature ρ of the beam, equal to CG or CG'.

Let us denote by λ the wavelength in the centre of the beam, shown in Fig. 5 by GG'. Then, if $2t$ is the thickness AB of the beam, we may write

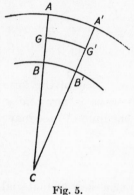

$$AA' = \lambda + \frac{\partial \lambda}{\partial n} t,$$

where $\partial/\partial n$ means differentiation normal to the beam. Then applying the rules of similar triangles to the triangles CGG', CAA', we see that

$$\frac{1}{\rho} = \frac{1}{\lambda} \frac{\partial \lambda}{\partial n}.$$

This formula will be used further in Chapter II, § 3.

Fig. 5.

3. DISPERSION, GROUP VELOCITY AND WAVE GROUPS

In wave motion of the type described by equation (6) there is no dispersion; that is to say the wave velocity v is independent of frequency. If, on the other hand, v is a function of frequency, dispersion will occur.

We require first to prove the formula for the group velocity v_G with which a group of waves propagates itself. This formula is

$$v_G = \frac{d\omega}{dk}, \tag{11}$$

where $\omega = 2\pi\nu$ and $k = 2\pi/\lambda$. The formula is often written

$$v_G = \frac{d\nu}{d(1/\lambda)}.$$

It will be noticed that since $\omega = vk$, for non-dispersive systems the wave and group velocities are the same.

The simplest way of obtaining formula (11) is to superimpose *two* simple harmonic waves with wave numbers k, k' differing by a small quantity. We have then for the resultant amplitude

$$\sin(kx - \omega t) + \sin(k'x - \omega't),$$

which is equal to

$$2 \sin\left(\frac{k+k'}{2}x - \frac{\omega+\omega'}{2}t\right) \cos\left(\frac{k-k'}{2}x - \frac{\omega-\omega'}{2}t\right).$$

This function represents a series of wave groups, as shown in Fig. 6a, each of length $\Delta x = 2\pi/(k-k')$, moving from left to right with velocity $(\omega-\omega')/(k-k')$. Since $k-k'$ and $\omega-\omega'$ are small, this may be written $d\omega/dk$.

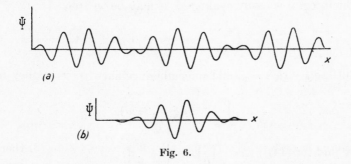

Fig. 6.

Wave groups of length Δx have thus been obtained by super-imposing two simple harmonic waves for which the values of k are separated by Δk, where

$$\Delta x \, \Delta k = 2\pi. \tag{12}$$

It is of interest to prove a similar theorem for a single wave group. Suppose that we add together simple harmonic waves with wave-numbers lying in the range Δk by setting for the wave amplitude

$$\Psi = \int_{-\infty}^{\infty} A(k)\, e^{ikx} dk, \tag{13}$$

where $A(k)$ vanishes, or tends rapidly to zero, outside a range Δk about the value k_0. It is convenient to take for $A(k)$ the Gaussian function, a form which permits the integrations to be carried out. We set

$$A(k) = \exp\{-(k-k_0)^2/(\tfrac{1}{2}\Delta k)^2\}.$$

The integral then becomes

$$\int_{-\infty}^{\infty} e^{-ak^2+bk+c} dk, \tag{14}$$

where
$$a = 1/(\tfrac{1}{2}\Delta k)^2,$$
$$b = ix + 2k_0/(\tfrac{1}{2}\Delta k)^2,$$
$$c = -k_0^2/(\tfrac{1}{2}\Delta k)^2.$$

The integral is easily evaluated; it may be written

$$\int_{-\infty}^{\infty} e^{-a(k-\frac{1}{2}b/a)^2+c+b^2/4a} dk.$$

Putting $k - \tfrac{1}{2}b/a = \xi$, and remembering that

$$\int_{-\infty}^{\infty} e^{-a\xi^2} d\xi = \sqrt{(\pi/a)},$$

we find for (14)
$$\sqrt{\left(\frac{\pi}{a}\right)} \exp\left(c + \frac{b^2}{4a}\right),$$

or, in our case, $\quad \tfrac{1}{2}\pi^{\frac{1}{2}} \Delta k \exp\{-(\tfrac{1}{4}\Delta k)^2 x^2 + ik_0 x\}.$

This represents a wave group of the form (Fig. 6b)

$$\text{const. } \exp\{-(x/\tfrac{1}{2}\Delta x)^2 + ik_0 x\},$$

where, equating $\tfrac{1}{4}\Delta k$ to $1/\tfrac{1}{2}\Delta x$, we have

$$\Delta x \Delta k = 8. \tag{15}$$

The somewhat different interpretations given to Δx, Δk account for the difference between (12) and (15).

It is of interest to generalise the integral (13) to investigate motion of the wave group, and thus to obtain the formula for the group velocity for a single group. At a subsequent time t the amplitude Ψ will be represented by

$$\Psi = \int_{-\infty}^{\infty} A(k) e^{ikx - i\omega t} dk.$$

If we expand

$$\omega = \omega(k_0) + (k - k_0)\left(\frac{d\omega}{dk}\right)_{k=k_0} + \tfrac{1}{2}(k - k_0)^2 \left(\frac{d^2\omega}{dk^2}\right)_{k=k_0},$$

then the integral may be evaluated as before. The reader will easily verify† that the wave group moves with velocity $d\omega/dk$, and that the width Δx increases, and becomes for large t

$$t\,\frac{d^2\omega}{dk^2}\,\Delta k.$$

4. FOURIER'S THEOREM AND CHARACTERISTIC FUNCTIONS

It will be convenient to introduce the reader to the problem of characteristic functions by writing down again the differential equation of a vibrating string (cf. equation (6))

$$\frac{\partial^2 \Psi}{\partial t^2} = v^2 \frac{\partial^2 \Psi}{\partial x^2}, \tag{16}$$

where Ψ is the displacement at a point distant x from one end of the string. We consider in this section the possible vibrations when the string is rigidly held at the two ends, for example at $x = 0$ and $x = a$.

We define as a 'normal mode' a mode of vibration of the string in which each point executes simple harmonic motion with frequency $\omega/2\pi$, say. Thus for a normal mode

$$\Psi(x,t) = \psi(x)\{A \cos \omega t + B \sin \omega t\}, \tag{17}$$

where $\psi(x)$ satisfies the equation,

$$\frac{d^2\psi}{dx^2} + \frac{\omega^2}{v^2}\psi = 0. \tag{18}$$

If the string is uniform, so that v is not a function of x, the solutions which vanish at $x = 0$ are

$$\psi = \sin(\omega x/v). \tag{19}$$

† Cf. N. F. Mott and H. S. W. Massey, *Theory of Atomic Collisions*, 1949, p. 17.

The displacement of the string must also vanish at $x = a$; this will be the case if

$$a\omega/v = n\pi, \tag{20}$$

where n is an integer. These values of ω, then, determine the frequencies with which the string can vibrate. Substituting (19) and (20) into (17), we see that the most general type of vibration of which the string is capable is that which results when all the normal modes are superimposed, namely,

$$\sum_n \sin(\pi n x/a)\{A_n \cos \omega_n t + B_n \sin \omega_n t\}, \tag{21}$$

where A_n, B_n are arbitrary constants and

$$\omega_n = n\pi v/a.$$

We shall now show how to determine the subsequent motion if the string is given any arbitrary initial displacement and released from rest. At time $t = 0$, then, let the displacement be

$$y = F(x),$$

where F is some function that vanishes at $x = 0$ and $x = a$. Since (21) is the general solution of the differential equation (16), that is to say, it represents the most general motion possible, it must be possible to represent the subsequent motion by a series of this type. Since the string is released from rest at time $t = 0$, the coefficients B_n must all vanish; the subsequent motion is thus given by

$$y = \sum_n A_n \cos \omega_n t \sin\left(\frac{\pi n x}{a}\right).$$

It follows that, putting $t = 0$, it must be possible to represent the function $F(x)$ by an expansion of the type

$$F(x) = \sum_n A_n \sin\left(\frac{\pi n x}{a}\right). \tag{22}$$

An expansion of this type is known as a Fourier series.

The coefficients A_n may be determined by making use of the orthogonal relation

$$\int_0^a \sin\left(\frac{\pi n x}{a}\right) \sin\left(\frac{\pi m x}{a}\right) dx = 0, \quad m \neq n,$$

which is easily verified. It may also be seen that

$$\int_0^a \sin^2\left(\frac{\pi n x}{a}\right) dx = \tfrac{1}{2}a.$$

If then we multiply both sides of (22) by $\sin(\pi m x/a)$ and integrate from 0 to a, all terms vanish except that in m; we have, therefore,

$$\tfrac{1}{2}a A_m = \int_0^a F(x) \sin\left(\frac{\pi m x}{a}\right) dx.$$

This equation determines the coefficients and hence the subsequent motion.

Exercises

(1) Determine the coefficients A_n for a string plucked at the centre, so that the initial displacement is given by

$$F(x) = bx \qquad 0 < x < \tfrac{1}{2}a,$$
$$= b(a-x) \quad \tfrac{1}{2}a < x < a.$$

Find the energy of each normal mode, and verify that the sum of the energies of all the normal modes is equal to the work done in displacing the string in the first place.

(2) A string of mass per unit length ρ, tension T and length $2a$, is rigidly fixed at the two ends $x = \pm a$. It is set in vibration by a sound wave which exerts on it a force $p \cos \omega t$ per unit length. Write down the equation of motion of the string, and verify that the solution corresponding to forced vibrations is

$$y = \frac{p}{\rho \omega^2}\left[1 - \frac{\cos(\omega x/v)}{\cos(\omega a/v)}\right] \cos \omega t,$$

where y is the displacement.

Write down also the complete solution of the equations, and, without working out any integrals, find the solution appropriate to the case where the string is at rest and undisplaced at time $t = 0$.

The Fourier expansion is a particular case of a more general type of expansion, which may be illustrated by considering the normal modes of a vibrating string of which the mass per unit

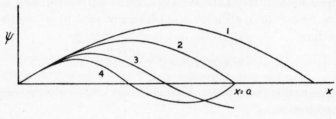

Fig. 7.

length (ρ) is not constant. It will be remembered that $v^2 = T/\rho$; v^2 will thus be a function of x, and we may write $1/v^2 = f(x)$, so that (18) becomes

$$\frac{d^2\psi}{dx^2} + \omega^2 f(x)\,\psi = 0. \tag{23}$$

Such an equation, together with the boundary conditions that ψ must vanish at $x = 0$ and $x = a$, defines a series of values of ω. For suppose we choose a small value of ω and obtain the integral of equation (23) that vanishes at $x = 0$; the solution, oscillating on account of the considerations of § 1, will cross the x-axis again at some value of x greater than a. As ω is gradually increased, a function ψ will be obtained which does vanish at $x = a$, as shown in Fig. 7. We call this the first characteristic function, and denote it by $\psi_1(x)$, and the corresponding value of ω by ω_1. Similarly $\psi_2(x)$ denotes the solution with one zero between $x = 0$ and $x = a$, and so on.

Exercises

(1) Prove the orthogonal relation

$$\int_0^a \psi_n(x)\,\psi_m(x)f(x)\,dx = 0$$

for two characteristic solutions of (23), and hence show how to expand any arbitrary function in a series of characteristic functions.

(2) Use the W.K.B. method to determine the values of ω_n corresponding to solutions of (23) for large values of n.

5. LEGENDRE POLYNOMIALS

The Legendre polynomials $P_l(\cos\theta)$ are defined as follows: If $|x| < 1$ we may expand the quantity $(1 - 2x\cos\theta + x^2)^{-\frac{1}{2}}$ in ascending powers of x. The polynomials are defined as the coefficients in the expansion. Thus

$$\frac{1}{\sqrt{(1 - 2x\cos\theta + x^2)}} = 1 + xP_1(\cos\theta) + x^2P_2(\cos\theta) + \dots.$$

It will easily be verified that

$$P_1(\cos\theta) = \cos\theta,$$

$$P_2(\cos\theta) = \tfrac{3}{2}\cos^2\theta - \tfrac{1}{2}.$$

The functions are orthogonal:

$$\int_0^\pi P_l(\cos\theta)\,P_{l'}(\cos\theta)\sin\theta\,d\theta = 0.$$

The importance of the Legendre functions is that, in spherical polar coordinates (r, θ, ϕ), the general solution of an equation of the type

$$\nabla^2\psi + F(r)\,\psi = 0, \tag{24}$$

which is a function of r, θ alone, is

$$P_l(\cos\theta)f(r),$$

where $f(r)$ satisfies

$$\frac{d^2f}{dr^2} + \frac{2}{r}\frac{df}{dr} + \left\{ F(r) - \frac{l(l+1)}{r^2} \right\} f = 0. \tag{25}$$

The reader will easily verify this for $l = 0$, $l = 1$.

To obtain the most general solution which is a function also of ϕ one has to introduce the associated Legendre polynomials, $P_l^u (\cos \theta)$. where u has the values $0, 1, 2, ..., l$. Some values are

$$l$$

u	0	1	2
0	1	$\cos \theta$	$\frac{3}{2}\cos^2 \theta - \frac{1}{2}$
1	—	$\sin \theta$	$3\cos \theta \sin \theta$
2	—	—	$3\sin^2 \theta$

The general solution of (24) is

$$P_l^{|u|}(\cos \theta)\, e^{iu\phi} f(r),$$

where f still satisfies (25), and u has $2l + 1$ integral values from $-l$ to $+l$.

CHAPTER II

THE WAVE EQUATION OF SCHRÖDINGER

1. THE WAVE FUNCTION

Wave mechanics is a system of equations which determines the behaviour of the fundamental particles of physics, the electron, the proton, the neutron, and their interaction with radiation. In its present form it appears adequate to describe the behaviour of the electrons outside the atomic nucleus sufficiently accurately to account for the known facts of spectroscopy, chemistry, and so on. Within the nucleus it has had some success, notably in giving an explanation of α-decay; but at the time of writing not sufficient is known about the forces between the constituents of the nucleus for a forecast to be made of its ultimate success in this field.

In this chapter we shall limit ourselves to the application of wave mechanics to electrons. The theory will be based on a single experimental fact, the diffraction of electron beams. This was first discovered by Davisson and Germer† and by G. P. Thomson,‡ and has now become a useful technique of applied physics. Briefly the experimental facts are as follows. When a beam of electrons passes through a crystalline substance, such as a metal foil, the beam is scattered by the substance in exactly the same way as a beam of X-rays is scattered. Thus diffraction rings are produced by a beam which has penetrated a polycrystalline foil; and from a single crystal a beam of electrons is reflected according to the Bragg law. The beam of electrons thus behaves as though it were a beam of waves, and the wavelength can be determined; it is related to

---

† C. Davisson and L. H. Germer, *Phys. Rev.* xxx, 707, 1927.
‡ G. P. Thomson, *Proc. Roy. Soc.* A, cxvii, 600, 1928.

the momentum p of each electron by the equation

$$\lambda = h/p, \tag{1}$$

where h is Planck's constant.[†]

In this book we prefer to treat (1) as given by experiment, and thus as the observed fact on which the whole theory of wave mechanics must be based. It was predicted, however, by Louis de Broglie[‡] before it was discovered experimentally. A simplified version of his argument is as follows:

If there is some relationship between the momentum vector **p** of the particle and some property of a train of waves, it must be obtained by equating **p** to some vector which describes the wave motion. Now a train of waves, travelling in the direction defined by the direction cosines (l, m, n), may be written

$$\sin\{k(lx + my + nz) - \omega t\}.$$

k is here the wave number, or 2π multiplied by the reciprocal of the wavelength. Since $k(lx + my + nz)$ is a scalar, and (x, y, z) is a vector, it follows that the quantity with components

$$(kl, km, kn)$$

is itself a vector. This we call the wave vector, and denote it by **k**. Thus if a correspondence of the type envisaged exists between the momentum of a particle and the wavelength of a wave, it must be of the form

$$\mathbf{p} = \text{const.}\,\mathbf{k}.$$

That the constant should be Planck's constant divided by 2π was suggested by the existence of a similar relationship for light quanta, where the momentum was known to be $h\nu/c$, in other words $hk/2\pi$.

Returning now to beams of electrons, one can define more precisely the observed behaviour as follows. In any problem in

[†] For experimental proof that beams of particles of atomic mass behave in the same way, cf. F. Knauer and O. Stern, Z. *Phys.* LIII, 786, 1929, or I. Estermann and O. Stern, Z. *Phys.* LXI, 115, 1930.

[‡] L. de Broglie, *Phil. Mag.* XLVII, 446, 1924; *Ann. Phys.*, Paris, III, 22, 1925.

which it is desired to calculate the path of a beam of electrons, its scattering by atoms or crystals or its bending by electric or magnetic fields, one has to postulate the presence of a wave, and calculate the wave amplitude everywhere. Then the density of electrons at any point will be proportional to the intensity of the wave at this point. All this is merely the expression of an experimental fact, the diffraction of electrons by crystals.

We denote the amplitude of this wave by Ψ. Since no way of measuring its amplitude exists except through the property that the intensity is proportional to the density of electrons, it is reasonable to choose our 'units' so that the square of the modulus of Ψ is *equal* to the density of electrons; thus

$$|\Psi|^2 = N,$$

where N is the number of electrons per unit volume.

The quantity Ψ, known as the wave function, is a complex quantity

$$\Psi = f + ig. \tag{2}$$

The square of the modulus is thus defined by

$$|\Psi|^2 = \Psi\Psi^* = f^2 + g^2.$$

The asterisk is used throughout this book to denote the complex conjugate of a complex quantity. Thus if

$$\Psi = f + ig,$$

then $\quad\quad\quad\quad \Psi^* = f - ig.$

It is often a stumbling-block to the beginner in this subject that a physical quantity, the wave function, should be represented by a complex quantity. The reason is as follows. We know *a priori* nothing about the wave function, but we should expect, by analogy with the case of light waves, that the type of expression which in other wave systems represents the energy density would in this case give the particle density.

But the energy density in any wave system is always given by the sum of the squares of two independent quantities whose magnitudes define the state of the wave. For a light wave these quantities are E and H and the energy density is $(E^2 + H^2)/8\pi$. For elastic waves they are the displacement and velocity of the medium. Thus, for the waves associated with electrons or other material particles, it is reasonable to assume that the state of the wave at any point is defined by two quantities f and g, and it is convenient to combine them into a single function Ψ by means of (2).

It may be noted that Maxwell's equations for the electromagnetic field in free space may be treated in the same way; the equations are

$$-c \operatorname{curl} \mathbf{E} = \frac{\partial \mathbf{H}}{\partial t}, \quad c \operatorname{curl} \mathbf{H} = \frac{\partial \mathbf{E}}{\partial t},$$

and if Ψ is written for $\mathbf{E} + i\mathbf{H}$, both equations may be combined in the single equation

$$c \operatorname{curl} \Psi = i\frac{\partial \Psi}{\partial t}.$$

It will be convenient at this stage to make a further assumption about the form of a plane wave. A plane wave travelling, say, along the x-axis has, for any type of wave motion, the form

$$A \sin(kx - \omega t + \epsilon),$$

where k is the wave number, $\omega/2\pi$ the frequency, and ϵ a phase. In a plane polarised light wave $\mathbf{E}$ and $\mathbf{H}$ are in phase; thus the energy density is proportional to

$$\sin^2(kx - \omega t + \epsilon)$$

and fluctuates with time at any point. There is no reason to think that any such fluctuation occurs in the wave associated with an electron; it would in fact be difficult to understand what physical significance could be ascribed to a rapid fluctuation of the probability that an electron would be found at a certain point. It is therefore reasonable to suppose that f and g are 90° out of phase, so that, A being a constant or a slowly varying function of x,

$$f = A \cos(kx - \omega t),$$

$$g = A \sin(kx - \omega t),$$

and
$$|\Psi|^2 = f^2 + g^2 = A^2.$$

With this convention $|\Psi|^2$ keeps a steady value independent of time. Making use of the complex function Ψ we see that a plane wave moving from left to right is represented by

$$\Psi = A e^{i(kx - \omega t)}.$$

We shall represent a wave going in the opposite direction by

$$\Psi = A e^{i(-kx - \omega t)}.$$

In the remainder of this book we shall follow the accepted convention and use always the complex wave function Ψ, and shall not refer again to the real and imaginary parts, f and g.

2. SCHRÖDINGER'S EQUATION

We shall now write down Schrödinger's wave equation in the form appropriate to a beam of electrons, each of total energy W, moving in an electrostatic field.

We represent by $V(x, y, z)$ the potential energy of an electron in this field; thus in a uniform electrostatic field E in the z direction, for example, we should have

$$V(x, y, z) = Eez;$$

for an electron in the field of a nucleus of charge Ze,

$$V(x, y, z) = -Ze^2/r,$$

where $r = \sqrt{(x^2 + y^2 + z^2)}$ is the distance from the nucleus. W, the total energy, is equal to the kinetic energy at the point where we arbitrarily choose $V(x, y, z)$ to be zero, at $z = 0$ for the first case and $r = \infty$ for the second. Then at the point (x, y, z) the kinetic energy of one of the particles is $W - V(x, y, z)$. Therefore the experimental relation (1) shows that the wavelength λ of the accompanying wave is

$$\lambda = h/\sqrt{\{2m(W - V)\}}. \tag{3}$$

Now it is clear from the considerations of Chapter I, § 2 that for motion in one dimension a function Ψ which oscillates in space with constant wavelength λ satisfies the equation

$$\frac{d^2\Psi}{dz^2} + \frac{4\pi^2}{\lambda^2}\Psi = 0,$$

and that the generalisation to three dimensions is

$$\nabla^2\Psi + \frac{4\pi^2}{\lambda^2}\Psi = 0.$$

The simplest assumption that we can make is that the same equation is satisfied where λ varies from point to point. Thus substituting from (3) we obtain

$$\nabla^2\Psi + \frac{8\pi^2 m}{h^2}(W - V)\Psi = 0. \tag{4}$$

This is Schrödinger's equation for the wave function Ψ.

It is convenient to introduce the symbol $\hbar$ to denote $h/2\pi$. With this notation the wave equation becomes

$$\nabla^2\Psi + \frac{2m}{\hbar^2}(W - V)\Psi = 0.$$

Certain results of this equation must be verified before it can be regarded as satisfactory. It must be shown:

(i) That it makes correct predictions about the bending of beams in electric and magnetic fields, where the classical Newtonian mechanics is known to give correct results.

(ii) That it predicts that the total current in a steady beam does not vary from point to point, so that the equation does not predict the creation or annihilation of particles.

We shall attend to these points in turn.

3. BENDING OF A BEAM OF ELECTRONS

According to Newtonian mechanics, the bending of a beam by an electric field can be calculated as follows. The force on each particle normal to the beam is $-\partial V/\partial n$, where $\partial/\partial n$ denotes differentiation normal to the beam. The radius of curvature R of the orbit of each particle is obtained by equating this to the centrifugal force, mv^2/R, which may be written $2(W-V)/R$. Thus

$$\frac{1}{R} = \frac{\partial V}{\partial n} \Big/ 2(W-V). \tag{5}$$

We wish to show that the same formula for the bending of a beam of electrons can be obtained by means of wave mechanics. We have already shown that if λ, the wavelength, is a function of position, the radius of curvature of the beam is given by (cf. Chap. I, § 2):

$$\frac{1}{R} = \frac{1}{\lambda}\frac{\partial \lambda}{\partial n}.$$

Since $\lambda = h/\sqrt{\{2m(W-V)\}}$, we see that formula (5) follows. Thus, in so far as the effect of an electric field is concerned, classical and quantum mechanics give the same result.

A similar result may be obtained for a magnetic field using (13), but the proof will not be reproduced here.

4. SOLUTIONS OF SCHRÖDINGER'S EQUATION AND THE CONSERVATION OF THE NUMBER OF PARTICLES

In the next chapter it will be shown that the frequency ν of an electron wave is related to the total energy W of the electron which it represents by the equation

$$W = h\nu.$$

Thus if $\psi(x,y,z)$ is any solution of (4) describing the behaviour of a beam of particles each of energy W, the full form of the wave function is

$$\Psi(x,y,z;t) = \psi(x,y,z)\,e^{-2\pi i \nu t}, \quad \nu = W/h. \tag{6}$$

The use of the negative sign in the exponential is simply a convention. As long as we are dealing with steady beams of particles all having the same energy, the time will enter into the wave function through a factor $e^{-2\pi i W t/h}$ as in (6). It will thus simplify all formulae dealing with such beams if we write down the functions ψ and not the functions Ψ; a plane wave going from left to right, for instance, will be written $\psi = e^{ikx}$.

With this simplification, certain examples will be considered which illustrate the novel features of wave mechanics and also verify the conservation of particles. We shall begin by considering the motion along the x-axis of a beam of electrons of infinite width in a field of potential energy $V(x)$. The Schrödinger equation then becomes

$$\frac{d^2\psi}{dx^2} + \frac{2m}{\hbar^2}\{W - V(x)\}\psi = 0.$$

In the absence of a field ($V = 0$), the solution representing a beam moving from left to right is

$$\psi = Ae^{ikx} \quad (k^2 = 2mW/\hbar^2),$$

which represents A^2 electrons per unit volume, or

$$A^2 v \quad (\tfrac{1}{2}mv^2 = W)$$

crossing unit area per unit time. In the presence of a field we may distinguish two cases:

(i) $V(x)$ varies slowly from point to point. An approximate solution may then be obtained by the W.K.B. method (Chap. I, § 1) and is

$$\psi = \frac{A}{(W - V)^{\frac{1}{4}}} \exp\left[i \int^x \left\{\frac{2m}{\hbar^2}(W - V)\right\}^{\frac{1}{2}} dx\right], \tag{7}$$

where A is a constant, and the lower limit of the integral is arbitrary. The form of the solution is illustrated in Fig. 8 for the case where $V = -eEx$, and thus for an electron accelerated by an electric field. It will be seen that as the electron is accelerated, so that the wavelength shortens, the amplitude also decreases. The number of electrons per unit volume is, by (7),

$$|\Psi|^2 = \frac{A^2}{(W-V)^{\frac{1}{2}}}.$$

But $W - V$ is the kinetic energy, so that $|\Psi|^2$ is inversely proportional to v, the velocity of the particles at the point con-

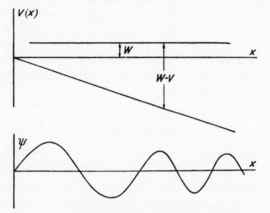

Fig. 8. Potential energy function and wave function ψ for an electron accelerated by a field.

sidered. Thus $v|\Psi|^2$, the number of particles crossing unit area per unit time, is the same at all points of the beam. The conservation of particles is thus verified.

If $V(x)$ is a slowly varying function of x, then the predictions made by wave mechanics are the same as those of classical mechanics; the electrons are accelerated by the field, and none of them is reflected. Here, as in the bent beam treated in § 3, wave mechanics makes no new predictions. If, however, $V(x)$ varies significantly in a distance small compared with the wavelength, the predictions of wave mechanics *are* entirely different from those of classical mechanics. This case will now be treated.

(ii) We may consider an extreme case, a 'potential jump', or in other words a plane perpendicular to the x-axis at which the potential energy function $V(x)$ changes discontinuously. This example is introduced in order to illustrate the principles of wave mechanics; no case of a discontinuous potential jump exists in nature. The example closest to that discussed here is perhaps the rapid change in the potential energy function which exists at the surface of a metal (cf. Chap. V, § 7).

We set then for $V(x)$

$$V(x) = 0 \quad x < 0,$$
$$= V_0 \quad x > 0,$$

and consider a stream of particles each of kinetic energy $W(W > V_0)$ incident on the potential jump from the left. At the potential jump there is a sudden change of wavelength; the wave-number changes from

$$k = \sqrt{(2mW)}/\hbar \qquad x < 0$$

to
$$k' = \sqrt{\{2m(W - V_0)\}}/\hbar \quad x > 0.$$

Therefore according to the arguments of Chapter I, § 2, the wave must be partially transmitted and partially reflected. In order to calculate how much is reflected, and how much transmitted, it is necessary to know the boundary conditions satisfied by the wave function. These are that ψ and $d\psi/dx$ are continuous. This may easily be seen, since

$$\frac{d\psi}{dx} = - \int^x \frac{2m}{\hbar^2} (W - V) \psi \, dx,$$

and, although the integrand is discontinuous, the integral (which represents the area under a curve) must be continuous.

With these boundary conditions, the analysis of Chapter I, § 2, may be applied as it stands. With an incident wave of amplitude unity (e^{ikx}), the amplitude of the reflected wave is $(k - k')/(k + k')$, and that of the transmitted wave is $2k/(k + k')$. The numbers of particles incident, reflected and transmitted per unit area per unit time are

$$v \qquad\qquad \text{incident,}$$
$$v(k - k')^2/(k + k')^2 \quad \text{reflected,}$$
and
$$4v'k^2/(k + k')^2 \quad \text{transmitted.}$$

The proportion R reflected is thus

$$R = (k - k')^2/(k + k')^2,$$

and the proportion T transmitted is

$$T = 4kk'/(k + k')^2.$$

It will easily be verified that $T + R$ is equal to unity. It is thus verified again that the wave equation chosen is compatible with the conservation of charge.

The prediction made by wave mechanics, that some of the particles are transmitted and some reflected, is of course fully at variance with classical mechanics, according to which they would all be transmitted. We see then that wave mechanics is unable to make a definite statement about the behaviour of an electron incident on a potential jump; it only allows a calculation of the average numbers transmitted and reflected, or in other words the probability that a given electron is transmitted or reflected. This inability to make exact predictions about the behaviour of individual particles is a general property of wave mechanics.

5. THE CURRENT VECTOR

The student will easily verify that, for a wave function of the type

$$\psi = Ae^{ikx} + Be^{-ikx},$$

the number of electrons $v(|A|^2 - |B|^2)$ crossing unit area per unit time may be written

$$\frac{\hbar}{2mi}\left(\psi^* \frac{\partial \psi}{\partial x} - \psi \frac{\partial \psi^*}{\partial x}\right).$$

A general proof that this quantity is independent of x is of interest. Since ψ and ψ^* satisfy the equations

$$\frac{d^2\psi}{dx^2} + \frac{2m}{\hbar^2}(W - V)\psi = 0$$

and

$$\frac{d^2\psi^*}{dx^2} + \frac{2m}{\hbar^2}(W - V)\psi^* = 0,$$

we have, multiplying the first equation by ψ^* and the second by ψ and subtracting,

$$\psi^* \frac{d^2\psi}{dx^2} - \psi \frac{d^2\psi^*}{dx^2} = 0.$$

In other words

$$\frac{d}{dx}\left(\psi^* \frac{d\psi}{dx} - \psi \frac{d\psi^*}{dx}\right) = 0. \qquad (8)$$

It follows that the current is independent of x. If this were not a consequence of the wave equation, the equation would lead to incorrect results.

In three dimensions the vector

$$\mathbf{j} = \frac{\hbar}{2mi}\{\psi^* \operatorname{grad}\psi - \psi \operatorname{grad}\psi^*\},$$

represents the number of electrons per unit time crossing unit area perpendicular to itself, and is known as the current vector.

Exercise

Prove the theorem equivalent to (8), that

$$\operatorname{div}\mathbf{j} = 0.$$

6. THE TUNNEL EFFECT

We have not yet considered the description in wave mechanics of a beam of electrons entering a field which opposes their motion and eventually stops them and turns them back. To describe what happens, we shall consider a beam of electrons moving from left to right along the x-axis and at $x = 0$ entering a field E. The potential energy of an electron is then given by

$$V(x) = eEx;$$

the kinetic energy is $W - V$, so the electrons are stopped and turned back when $x = W/eE$.

To describe the behaviour of the electrons according to wave mechanics, we have to solve the Schrödinger equation

$$\frac{d^2\psi}{dx^2} + \frac{2m}{\hbar^2}\{W - V(x)\}\psi = 0,$$

for this form of $V(x)$. Consider first the form of the solution to the right of the point where $x = W/eE$. $W - V$ is then negative, and the arguments of Chapter I, § 1 show that there are two solutions, one of which increases with increasing x and one of which decreases. The solution which represents the physical state of affairs is the one which decreases; the other solution would represent a rapidly increasing density of particles beyond W/eE, which is absurd. We choose then the decreasing solution; this is, of course, a real function of x.

To the left of $x = W/eE$, then, we have an oscillating solution $\psi(x)$, which, since it fits onto the real solution to the right, will be real too. Thus the complete wave function with the time factor

$$\Psi(x) = \psi(x)\, e^{-2\pi i\nu t}$$

represents a standing wave, that is to say, an incident wave and a reflected wave having equal amplitudes.

The description given by wave mechanics of the phenomenon is not very different from that given by classical mechanics; all the electrons are reflected. The only difference is that they are not all reflected exactly at the point where $x = W/Ee$; some of them penetrate a little further. There is thus a finite if small probability of finding an electron at any distance beyond the point where $x = W/Ee$.

This fact has important consequences; it gives to an electron a finite probability of penetrating through what is called in quantum mechanics a 'potential barrier'. A potential barrier is illustrated in Fig. 9b, formed by two fields; a field E opposing the electron's motion from $x = 0$ to $x = a$, and a field E in the opposite direction from $x = a$ to $x = 2a$. The potential energy of the electron in this field is

$$V(x) = Eex \qquad 0 < x < a,$$
$$V(x) = Ee(2a - x) \quad a < x < 2a.$$

In general a potential barrier is a region in which $V(x) > W$ sandwiched between two regions in which $W > V(x)$. According

3

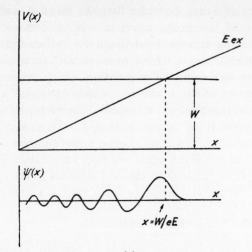

(a)

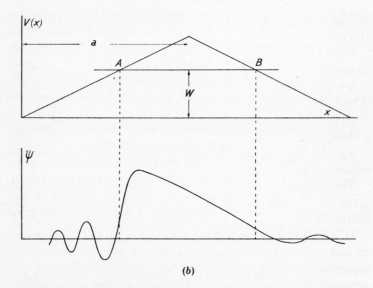

(b)

Fig. 9. (a) Electrons reflected by a field. (b) A potential barrier due to two fields as described. V is the potential energy of an electron and ψ the wave function.

to classical mechanics, particles incident on such a region will all be reflected; according to wave mechanics a certain proportion of the particles will penetrate the barrier and come out the other side. This prediction of wave mechanics is known in the literature as the 'tunnel effect'.

If ψ_A, ψ_B are the amplitudes of the wave function at the two extremities of the barrier, the chance of penetration P is given approximately by

$$P = |\psi_B/\psi_A|^2. \qquad (9)$$

In many practical cases P can be calculated by the W.K.B. method. Neglecting the factor outside the exponential in (7), which in general changes little in comparison with the exponential factor, we see from (9) that

$$P = \exp\left[-2\int_A^B \left\{\frac{2m}{\hbar^2}\left(V(x) - W\right)\right\}^{\frac{1}{2}} dx\right], \qquad (10)$$

the integration being from one extremity of the barrier to the other.

Some physical phenomena in which the tunnel effect is important are:

(a) The escape of α-particles from a radioactive nucleus (Chap. VII, § 3).

(b) The escape of electrons from a metal under the action of a strong field (Chap. V, § 7).

(c) The passage of electric current between two metals separated by an oxide layer.

The discussion of waves with imaginary refractive index given in Chapter I, § 2, is relevant to the phenomena considered here.

It is important to estimate how thick a barrier electrons can in fact penetrate. Let us suppose that two metal wires are separated by an air gap or an oxide layer of thickness a. If a potential difference of, say, half a volt is applied across the gap, a certain current will pass. Let us suppose that the area of the contact is 1 sq. mm. A metal contains about 10^{23} free electrons per c.c., and they move with a speed of about 10^8 cm./sec., so that 10^{29} will impinge on each side of the gap each second. If half a volt is applied, about one-tenth of the mean kinetic energy of the electrons (cf. Chap. VI, § 7), we may suppose that one-tenth of the electrons have not energy enough to pass through the gap against the field. Thus the number of electrons passing through the gap in the direction of the field is, each second, about

$$10^{28} P,$$

where P is the quantity defined by (9). A current of one ampere corresponds to c. 10^{19} electrons per sec., so we see that a current of this magnitude will flow if $P = 10^{-9}$, while a milliampere will flow if $P = 10^{-12}$. Taking for the height ϕ of the barrier a quantity of the order of the work function,

$$\phi \sim 4eV,$$

we have, since

$$\exp\{-2\sqrt{(2m\phi)}\,a/\hbar\} = P,$$

$$a = \frac{\hbar}{2\sqrt{(2m\phi)}} \ln\left(\frac{1}{P}\right) = 1{\cdot}13 \times 10^{-8} \log_{10}\left(\frac{1}{P}\right) \text{cm}.$$

Thus a gap of 10^{-7} cm. would give one ampere, $1{\cdot}4 \times 10^{-7}$ cm. a current of one milliampere. It will be seen that the current drops very rapidly as the thickness is increased, and barriers of c. 2×10^{-7} cm. are practically opaque.

7. SCATTERING OF BEAMS OF PARTICLES BY ATOMS

An important class of question which wave mechanics can solve is that of the scattering of beams of particles by atoms or nuclei. The problem can be put as follows. A substance, for example a gas, contains N scattering centres (atoms, nuclei) per unit volume. A particle (electron, proton, α-particle) moves through the substance. What is the probability, per length x of its path, that the particle is scattered through an angle θ into the solid angle $d\omega$? We denote this probability by

$$NI(\theta)\,x\,d\omega.$$

It is clear that $I(\theta)$ has the dimensions of an area. The integral

$$A = \int I(\theta)\, d\omega = \int_0^\pi I(\theta)\, 2\pi \sin\theta\, d\theta,$$

gives the effective total cross-section of the atom or other centre for the type of collision in question. In other words, NA is the chance of a collision per unit length of path.

By the methods developed in this chapter it is possible only to calculate the scattering of particles by a centre of force, for example the scattering of α-particles by the field of a heavy nucleus. The scattering of electrons by atoms is a many-body problem, involving the possibility of transfer of energy between the incident electron and the electrons of the atoms. It is, however, possible to consider approximately elastic collisions (those in which the electron loses no energy) by representing the atom as a centre of force. It will be shown in Chapter IV, § 6 how this force may be calculated. We denote by $V(r)$ the potential energy of an electron acted on by this force.

The problem, then, is to calculate $I(\theta)$ for a stream of particles incident on a region, at a distance r from the centre of which the potential energy of any one particle is $V(r)$. One has therefore to find a solution of Schrödinger's equation

$$\nabla^2 \psi + \frac{2m}{\hbar^2}(W - V)\psi = 0,$$

which represents an incident plane wave and a scattered wave. Such a solution must have the form, for large r,

$$\psi \sim e^{ikx} + \frac{e^{ikr}}{r} f(\theta).$$

Here the first term represents the incident wave moving from left to right along the x-axis, the second the scattered wave which must fall off inversely as r. The solution represents a state of affairs in which v particles cross unit area per unit time in the incident beam; in the scattered beam there are $r^{-2}\,|f(\theta)|^2$ particles per unit volume at a distance r from the scattering centre. Thus $v|f(\theta)|^2 d\omega$ cross an area $r^2 d\omega$ per unit time; hence

$$I(\theta) = |f(\theta)|^2.$$

One of the most important cases is that in which $V = ZZ'e^2/r$, corresponding to the scattering of particles of charge $Z'e(Z' = 2$ for α-particles, $Z' = -1$ for electrons) by a bare nucleus of charge Ze. Here an exact analysis† shows that

$$|f(\theta)|^2 = \left(\frac{ZZ'e^2}{2mv^2} \operatorname{cosec}^2 \tfrac{1}{2}\theta\right)^2.$$

It is remarkable that for this case—and for this case alone—wave mechanics yields the same formula as classical mechanics. This formula was in fact derived from classical mechanics by Darwin‡ and used by Rutherford to interpret his experiments on the scattering of α-particles by metal foils which established the nuclear model of the atom.

For a more general field $V(r)$, it may be shown that each element of volume $dx\,dy\,dz$ scatters a wavelet of amplitude, at a distance R from the element,

$$\frac{1}{R} \frac{2\pi m}{h^2} V(r)\,dx\,dy\,dz \times \psi, \tag{11}$$

where ψ represents the amplitude of the whole wave at that point. This may be shown in two ways. The rigorous method is to make use of a theorem known as Green's theorem (outside the scope of this book); the proof is given by Mott and Massey.§ A more elementary method is as follows. Consider a beam of electrons moving along the x-axis and incident on a region of sheet-like form in which the potential is defined by

$$\begin{aligned} V &= 0 & x &< 0, \\ &= V_0 & 0 &< x < a, \\ &= 0 & a &< x, \end{aligned}$$

where a is some small distance. We shall suppose that $W > V_0$.

† W. Gordon, Z. *Phys.* xlviii, 180, 1928; Mott and Massey, chap. iii.
‡ C. G. Darwin, *Phil. Mag.* xxvii, 499, 1914.
§ Mott and Massey, chap. vi.

We may calculate the amplitude reflected as follows. We set

$$\psi = e^{ikx} + Ae^{-ikx} \qquad x < 0,$$
$$= Be^{ik'x} + Ce^{-ik'x} \qquad 0 < x < a,$$
$$= De^{ikx} \qquad a < x,$$

where
$$k^2 - k'^2 = 2mV_0/\hbar^2.$$

Putting in the boundary conditions that ψ and $d\psi/dx$ are continuous at $x = 0$ and $x = a$, and making also the assumption that $k'a \ll 1$, we find

$$A = \tfrac{1}{2}ia(k^2 - k'^2)/k.$$

Now our problem is to find the amplitude of a wavelet scattered by a volume element $dxdydz$ in which the potential is V_0. Let this be $\alpha dxdydz/R$. A surface element of the sheet of area dS will then scatter a wavelet $\alpha adS/R$. These wavelets add up to give a reflected wave of amplitude, at a point distant x from the sheet,

$$\alpha a \int_0^\infty \frac{e^{ikR}}{R} 2\pi z\, dz,$$

where $R^2 = z^2 + x^2$; on integration this gives, for $x < 0$,

$$2\pi\alpha aie^{-ikx}/k.$$

This has to be equated to Ae^{-ikx}. Thus

$$2\pi\alpha a/k = \tfrac{1}{2}a(k^2 - k'^2)/k,$$

whence
$$\alpha = (k^2 - k'^2)/4\pi = 2\pi mV_0/h^2,$$

which is what we set out to prove.

The result (11) may now be used to calculate the scattering of electron waves by a centre of force, if W is everywhere great compared with V. Under these conditions the form of the wave cannot be greatly perturbed within the atom, so that in calculating the scattered wavelets one can assume that ψ at any point is given by e^{ikx}. This approximate method of obtaining the scattered amplitude $f(\theta)$ is known as the Born† approximation. In fact, of course, near the centre of an atom $V(r)$ always becomes large; nevertheless, the Born approximation does give fair results for electrons of high energy. Making use of Born's approximation, the amplitude of the wave scattered through an angle θ may be calculated as follows. Let OXZ be

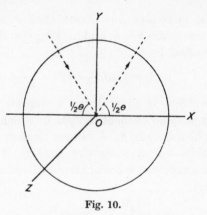

Fig. 10.

the plane normal to the bisector of the angle between the incident and reflected rays (Fig. 10). Then all the wavelets scattered from any plane parallel to OXZ will be in phase with each other, and the wavelets scattered from two such planes distant y from each other will have a phase difference μy, where, as may easily be verified,

$$\mu = 2k \sin \tfrac{1}{2}\theta, \quad k = 2\pi/\lambda.$$

The resultant, therefore, of all scattered wavelets will be, at large distances R from the atom, $f(\theta)/R$, where

$$f(\theta) = \frac{2\pi m}{h^2} \iint\!\int e^{i\mu y} V(r)\, dx\, dy\, dz.$$

† M. Born. Z. *Phys.* xxxvii, 863, 1926; Z. *Phys.* xxxviii, 803, 1926. The former paper is the first in which the probability interpretation of the wave function is introduced explicitly.

If we take spherical polar coordinates (r, θ', ϕ') such that $y = r \cos \theta'$, this becomes

$$f(\theta) = \frac{2\pi m}{h^2} \int_0^\pi \int_0^\infty e^{i\mu r \cos \theta'} V(r) r^2 dr \, 2\pi \sin \theta' d\theta'.$$

Writing $\cos \theta' = z$, $\sin \theta' d\theta' = - dz$, we find finally on integrating over z

$$f(\theta) = \frac{8\pi^2 m}{h^2} \int_0^\infty V(r) \frac{\sin \mu r}{\mu r} r^2 dr. \tag{12}$$

The integration may be carried out for various forms of $V(r)$.

Exercises

(1) Carry out the integration in (12) for the screened Coulomb field

$$V(r) = \frac{Ze^2}{r} \exp(-qr).$$

Show that as $q \to 0$, the Rutherford scattering formula is obtained.

(2) If $V(r)$ is the potential energy function due to a nucleus of charge Ze and a negative charge distribution of charge density $e\rho(r)$, show that (12) can be put in the form

$$f(\theta) = \frac{e^2}{2mv^2} \{Z - F(\theta)\} \operatorname{cosec}^2 \tfrac{1}{2}\theta,$$

where

$$F(\theta) = 4\pi \int_0^\infty \rho(r) \frac{\sin \mu r}{\mu r} r^2 dr.$$

Interpret this as showing that each element $d\rho$ of charge scatters according to the Rutherford law.

(3) From (12) show that the scattering is spherically symmetrical if the radius of the atom is small compared with $\lambda/2\pi$, where λ is the wavelength of the incident wave. Show that this is true, in general, of the *exact* solution of the wave equation.

(4) Find the scattering due to a hard sphere of radius a, on the surface of which ψ may be assumed to vanish; you may take $a \ll \lambda/2\pi$. Show that the total cross-section is $4\pi a^2$.

8. ELECTRONS IN A MAGNETIC FIELD

The Schrödinger equation for an electron moving in an electromagnetic field is†

$$\nabla^2\psi + \frac{2m}{\hbar^2}\left\{(W-V)\psi - \frac{\hbar e}{imc}(\mathbf{A}\,\mathrm{grad}\,\psi)\right\} = 0, \qquad (13)$$

where $\mathbf{A}$ is the vector potential of the field, defined by

$$H = \mathrm{curl}\,\mathbf{A}.$$

For a uniform field H along the x-axis, the vector potential is

$$\mathbf{A} = (0, -\tfrac{1}{2}Hz, \tfrac{1}{2}Hy),$$

so that the equation reduces to

$$\nabla^2\psi + \frac{2m}{\hbar^2}\left\{(W-V)\psi + \frac{\hbar eH}{2mci}\left(z\frac{\partial\psi}{\partial y} - y\frac{\partial\psi}{\partial z}\right)\right\} = 0.$$

Exercise

Using (13), prove (8) verifying the conservation of charge.

† For the proof, see, for example, N. F. Mott and I. N. Sneddon, *Wave Mechanics and its Applications*, 1948, p. 39.

CHAPTER III

WAVE GROUPS AND THE UNCERTAINTY PRINCIPLE

1. THE FREQUENCY OF ELECTRON WAVES

In the last chapter we have considered the description in terms of wave mechanics of beams of electrons, each electron of which has the same energy W. We have introduced a wave function Ψ of the form

$$\Psi = \psi(x, y, z) e^{-i\omega t}, \tag{1}$$

where ψ satisfies the equation

$$\nabla^2 \psi + \frac{2m}{\hbar^2}(W - V)\psi = 0, \tag{2}$$

and have interpreted the solution by saying that $|\Psi(x, y, z; t)|^2$ is the average density of electrons in the beam at the point (x, y, z), or that $|\Psi|^2 dx\,dy\,dz$ is the probability that an electron will be found at any moment in the volume element $dx\,dy\,dz$. We now show how to apply wave mechanics to a more general case than that of steady beams, namely, to a state of affairs where the density varies with the time. At the same time we shall introduce an expression for the frequency $\nu(= \omega/2\pi)$ of electron waves.

We consider first the following idealised experiment. Suppose that a beam of particles each having velocity v is incident on a screen, in which there is a hole which can be closed by a shutter (Fig. 11). The shutter is closed initially, then opened for a time t_0 and then closed again. Then a beam of length vt_0 will pass through the hole and move forward with velocity v. According to the concepts of wave mechanics, however, we must describe the whole phenomenon in terms of the wave function Ψ. A continuous train of waves falls on the screen;

when the shutter is opened a train of waves of limited length, that is to say a wave group or wave packet, is allowed to pass

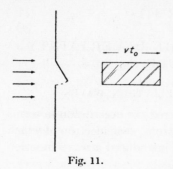

Fig. 11.

through. As usual, we must set the intensity of the wave at any point equal to the density of electrons there. The wave group must travel with the group velocity of the waves; thus if wave mechanics is to give a correct description of the observed phenomena, the group velocity of the waves must be equal to the actual velocity of the particles that they represent.

The group velocity in any type of wave motion is (cf. Chap. I, § 3) $d\omega/dk$, where $\omega(= 2\pi\nu)$ is defined by (1). We must thus set

$$\frac{d\omega}{dk} = v. \tag{3}$$

Up to the present we have not ascribed any physical meaning to the frequency ν of electron waves. By integrating (3), however, we may find an expression for ω and hence ν. For, by equation (1) of Chapter II, we see that

$$k = mv/\hbar.$$

Thus (3) becomes

$$\frac{d\omega}{dk} = \frac{\hbar k}{m}.$$

On integrating we find

$$\omega = \tfrac{1}{2}\hbar k^2/m + \text{const.}$$
$$= \tfrac{1}{2}mv^2/\hbar + \text{const.}$$

Thus for a freely moving particle we may set

$$\hbar\omega = h\nu = \text{kinetic energy} + \text{constant.}$$

We next have to consider the value of this constant. It will be realised that a steady beam of particles moving through a field of force must be represented by a wave with frequency the same at all points. It is natural to take a point where the potential energy is zero and to define $h\nu$ as the kinetic energy there. We see therefore that

$$\hbar\omega = h\nu = W, \tag{4}$$

where W is the *total* energy of each electron. It must be realised, however, that the point where the potential energy vanishes is quite arbitrary, and so the total energy of a particle in a field of force contains an arbitrary constant.

It is rather surprising that the frequency of these electron waves should also contain an arbitrary constant; it suggests that, though the equations of wave mechanics are correct in their description of how matter actually behaves, these waves have not the same sort of physical reality as sound or electromagnetic waves. This view will be confirmed by the considerations of Chapter V.

2. THE WAVE EQUATION FOR NON-STATIONARY PHENOMENA

The wave equation (2) applies to steady beams; it will not apply to wave groups such as that shown in Fig. 11. The equation that we require must contain terms in $\partial\Psi/\partial t$, so that it may be used to calculate the future motion of a wave group when its initial form is given. Moreover it must be of the first order in the time; that is to say it must contain terms in $\partial\Psi/\partial t$ but not in $\partial^2\Psi/\partial t^2$. This is because, as we have seen in Chapter II, the complex function Ψ contains two real terms f and g which are analogous to the displacement and the velocity of waves on a string, or to E and H in electromagnetic theory; thus a knowledge of Ψ alone at a given time, without a knowledge of $\partial\Psi/\partial t$, should suffice for the calculation of its value at all subsequent times. Only if the equation is of the first order will this initial condition be enough.

The required equation can be obtained by eliminating W from (2). From (4) we have for the typical wave function describing electrons of energy W

$$\Psi(x, y, z; t) = \psi(x, y, z)\, e^{-iWt/\hbar}. \tag{5}$$

Differentiating this equation, we obtain

$$\frac{\partial \Psi}{\partial t} = -\frac{iW}{\hbar}\Psi.$$

Substituting in (2), we find

$$\frac{\hbar}{i}\frac{\partial \Psi}{\partial t} = \frac{\hbar^2}{2m}\nabla^2\Psi - V\Psi. \tag{6}$$

This is the required equation. Its most general solution is made up by superimposing solutions of the type (5):

$$\Psi = \sum_W A_W \psi_W(x, y, z) e^{-iWt/\hbar}.$$

As will be shown in Chapter VI, such a solution represents a state of affairs in which the energy of the electron is not known, but the probability that it has the value W is $|A_W|^2$.

Exercise

Make use of (6) to verify the conservation of number of particles, namely, to prove that

$$\frac{\partial}{\partial t}\iiint |\Psi|^2 dx\,dy\,dz = 0,$$

the integral being over all space.

Notation

It is often convenient to write H for the operator,

$$H = -\frac{\hbar^2}{2m}\nabla^2 + V,$$

so that the Schrödinger equation (6) becomes

$$-\frac{\hbar}{i}\frac{\partial \Psi}{\partial t} = H\Psi,$$

and the equation (2)

$$(H - W)\psi = 0.$$

3. THE DESCRIPTION IN WAVE MECHANICS OF A SINGLE PARTICLE

The wave group described in § 1 and illustrated in Fig. 11 describes a number of electrons; the integral

$$\iiint |\Psi|^2 dx\, dy\, dz \tag{7}$$

gives the average or probable number of electrons passing through the hole while the shutter is open. It need not be an integer; and it may be less or greater than unity. We might, however, imagine the shutter open just long enough to let, on the average, one electron pass through; then the integral (7) will be set equal to unity, and the volume occupied by the wave group, the shaded area in Fig. 11, represents the space where, as a result of the experimental arrangement illustrated, the electron may be.

This arrangement with a shutter may be thought of as just a way of obtaining approximate information about the position and velocity of a particle. Many other devices may be imagined. Given any such device we may formulate as follows the way in which wave mechanics must be used to make predictions about the *future* position and velocity of the particle. We confine our description to movement in one dimension, though it may at once be generalised to three. Suppose that measurements are made, at a given instant of time, of the position and momentum of the particle. Suppose that the results of these measurements are that the position is at x_0 with a probable error $\pm \frac{1}{2}\Delta x$; and that the momentum is p_0 with a probable error $\pm \frac{1}{2}\Delta p$. Now if $\Delta x \Delta p$ is not too small, the result of these measurements can be described by a wave function having the form of a wave group. The chance that the particle is between the points x, $x + dx$ as the result of our measurement may be written

$$\text{const. } \exp\{-(x-x_0)^2/(\tfrac{1}{2}\Delta x)^2\}\, dx.$$

This by the rules of wave mechanics, is equal to $|\Psi|^2 dx$; we thus set at $t = 0$

$$\Psi = \text{const. } e^{ik_0 x} \exp\{-(x-x_0)^2/2(\tfrac{1}{2}\Delta x)^2\}, \tag{8}$$

where $k_0 = p_0/\hbar$. Such a wave function, then, represents a particle at the required position, and with momentum approximately equal to p_0. But (8) can be expanded, as shown in Chapter I, § 3, in the form

$$\psi = \int A(k) \exp\{ik(x - x_0)\}\, dk,$$

where $\qquad A(k) = \text{const.} \exp\{-(k-k_0)^2 (\tfrac{1}{2}\Delta x)^2/2\}.$

The wave group is thus made up of simple harmonic waves with k in a range about k_0 of $\pm \tfrac{1}{2}\Delta k$, where

$$\Delta k = 4\sqrt{2}/\Delta x,$$

as may indeed be seen without mathematical development from the considerations of Chapter I, § 3. The wave group thus describes particles with momenta p lying in the range determined by $|A(k)|^2$, and thus between $p_0 \pm \tfrac{1}{2}\Delta p$, where

$$\Delta p\, \Delta x = 4\hbar.$$

To a good enough approximation, this may be written

$$\Delta p\, \Delta x \sim \hbar. \qquad (9)$$

Equation (9) states the uncertainty principle of Heisenberg. If measurements are made so that $\Delta p\, \Delta x$ is greater than h, it is still possible to imagine a wave group set up, similar to a wave group of white light, containing waves having a range of frequencies; but if $\Delta x\, \Delta p$ is less than h, it is impossible to set up a wave group to represent the results of the measurements. We are thus driven to the conclusion that

either measurements for which $\Delta x\, \Delta p < h$ are impossible in the nature of things,

or it is impossible to describe the motion of an electron by means of wave mechanics.

The facts of electron diffraction seem to rule out the second alternative; we are thus driven to believe that there is in fact a limiting accuracy of all measurement. We shall come back to this point in the next section.

Once a wave group has been set up describing the results of the initial measurement, the wave equation (6), being linear in $\partial/\partial t$, will predict its form at any future time. Thus at any future time it is possible to state the probability $|\Psi(x,t)|^2 dx$ that the particle will be found between x and $x+dx$. This then is the type of prediction that wave mechanics enables one to make: given certain initial measurements, made with a certain probable error, one can predict the chance that at any future

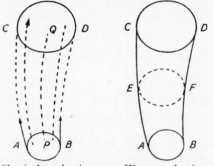

Classical mechanics   Wave mechanics

Fig. 12. Showing the contrast between the classical and wave-mechanical method of prediction. In the classical method a measurement shows that the particle is in the volume AB moving within the directions shown by the arrows. By considering all the orbits such as PQ consistent with this original measurement, one arrives at the conclusion that after time t the particle will be within the volume CD. The wave-mechanical treatment pictures a wave packet moving from AB to CD, passing through the intermediate position EF.

time a particle (or system of particles) will be found at a given point. In this, wave mechanics is similar to classical mechanics; but classical mechanics proceeds by calculating the system of orbits which are consistent with the original measurement; these are absent in wave mechanics. The difference between the two methods is illustrated in Fig. 12.

In fields which vary slowly with the distance it may be shown† that the wave group of wave mechanics follows the

† Cf., for example, Mott and Sneddon, chap. i.

4

same path as the particles of classical mechanics; it is only
when we have to deal with fields varying in a distance comparable with the wavelength of an electron (c. 10^{-8} cm.) that
the two systems give different results, as in the diffraction or
scattering of electrons by atoms.

4. THE UNCERTAINTY PRINCIPLE

We have seen that, if wave mechanics is valid, measurements must be impossible unless

$$\Delta p \, \Delta x > h.$$

It will be of interest to estimate the magnitude of these quantities. If we write $p = mv$,

$$\Delta v \, \Delta x > h/m$$
$$\sim 7 \text{ c.g.s. units for an electron.}$$

Thus if Δx is 1 cm., $\Delta v \sim 7$ cm./sec., which is of order one part in
10^8 of the velocity of electron in an atom. If, however, Δx is of
the order of the size of an atom (10^{-8} cm.), $\Delta v/v \sim 1$.

It is of great interest to examine the hypothetical experiments by which we could determine position and momentum
simultaneously, and to show that they do in fact yield an
uncertainty of the predicted amount. The most famous of
these demonstrations is the 'gamma-ray microscope' first discussed by Heisenberg. The argument put forward is as follows.
A beam of electrons is supposed to be travelling along the
x-axis with known momentum p. It is desired to observe an
electron and to measure its position; for this purpose it is
imagined that a microscope will be used, and since the utmost
resolving power is required a short wavelength should be chosen.
The position can then be determined to an accuracy given by

$$\Delta x = \lambda f/a,$$

where a is the aperture, λ the wavelength, and f the distance
from the electron to the lens.

Radiation cannot be scattered by an electron without disturbing the electron; radiation is scattered by free electrons according to the rules of the Compton effect, according to which the momentum lost by the light quantum when scattered is transferred to the electron. Thus if a quantum having frequency ν, and hence momentum $h\nu/c$, is scattered through an angle θ, momentum equal to

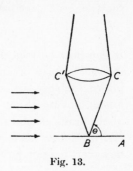

Fig. 13.

$$h\nu(1 - \cos\theta)/c$$

is transferred to the electron. Thus we cannot observe the electron without disturbing it. Moreover, we disturb it by an unknown amount, since, owing to the finite aperture of the lens, θ is not known exactly. In Fig. 13, θ may lie between ABC and ABC'. There is thus an uncertainty a/f in θ and hence, since $\theta \sim 90°$, of

$$h\nu a/cf$$

in the momentum transferred to the electron. Since $\lambda = c/\nu$ this may be written

$$\Delta p = ha/f\lambda,$$

where Δp is the uncertainty in the momentum of the electron after the measurement has been made. We see that

$$\Delta p\,\Delta x = h,$$

as we expect.

CHAPTER IV

STATIONARY STATES

1. THE OLD QUANTUM THEORY

It can now be regarded as an experimental fact that the total *internal* energy of an atom or molecule is quantised. By the internal energy we mean the total energy of the electrons and nuclei moving about their centre of gravity; the kinetic energy of the translational motion of the atom or molecule as a whole can of course have any positive value. The internal energy, however, cannot have any arbitrary value, but only one of a series of discrete values, of which one is the lowest. This is what is meant by the statement that the energy is quantised.

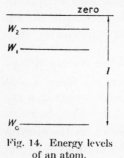

Fig. 14. Energy levels of an atom.

The simplest and most important application of this principle is to the energy of the electrons in an isolated atom, and thus in an atom of a monatomic gas or vapour. It is usual to measure this quantity with the convention that the total energy is zero when one electron is removed far from the atom and is at rest. With this convention, the quantised energy values of an atom are negative. A typical scheme is shown in Fig. 14. The distance of each horizontal line from the zero represents the energy of the atom in one of the quantised states. The lowest state of the atom is known as the normal or ground state, the higher states as excited states. The energy I required to remove an electron to a state at rest at infinity from an atom in the normal state is known as the ionisation potential.

No detailed review of the experimental evidence for the existence of quantised states in atoms will be given here, but we may mention the following:

(a) The existence of a definite energy required to excite an atom, and the fact that it is large compared with the thermal energies† of molecules in a gas, follow from the observation that the specific heat per gramme atom of a monatomic gas is $\frac{3}{2}R$, which can all be accounted for by translational motion. Collisions between gas atoms therefore do not change the internal energy.

(b) Many experiments‡ have been carried out which show that electrons after hitting an atom are deflected either without loss of energy, or with loss equal to one of the excitation potentials $W_1 - W_0$, $W_2 - W_0$ of Fig. 14 or else greater than the ionisation potential I.

(c) Detailed information about the energy levels is derived from spectroscopic evidence, coupled with the hypothesis that radiation of frequency ν is emitted and absorbed in quanta according to the equation

$$h\nu = W_n - W_m;$$

assuming this hypothesis, the existence of line spectra proves the existence of stationary states.

It should be emphasised that only the isolated atom in a gas or vapour has a system of energy levels of the type shown in Fig. 14. The electronic system of an isolated molecule has a similar system, but in addition the vibrational motion of the nuclei about their mean positions and the rotation of the molecule as a whole introduce additional series of levels, much closer together. The energy levels of electrons in solids are not quantised (cf. Chap. V, § 7).

The hypothesis of the existence of stationary states was introduced into physics by Niels Bohr§ in 1913. At the same time he introduced another hypothesis in order to be able to calculate the values of the quantised energies for the case of a

† The thermal energy $\frac{3}{2}kT$ of an atom at room temperature is 0·037 of an electron volt ($kT \sim 1/40$ eV.); the excitation potentials are of order 3–20 eV.

‡ Cf., for instance, E. G. Dymond and E. E. Watson, *Proc. Roy. Soc. A*, cxxii, 571, 1929.

§ N. Bohr, *Phil. Mag.* xxvi, 1, 476, and 857, 1913.

single particle moving round a centre of force, or a pair of particles moving round their centre of gravity. This hypothesis is as follows: the orbits are as predicted by Newtonian mechanics, but only those orbits are found in nature for which the total angular momentum is a multiple of $h/2\pi(=\hbar)$. This hypothesis, though extremely valuable at the time, has now been abandoned in favour of the description given by wave mechanics.

With the aid of this hypothesis one can show:[†]

(a) That the energy of an electron moving round a nucleus of charge Ze and of infinite mass is

$$W_n = -\frac{mZ^2e^4}{2\hbar^2}\frac{1}{n^2},$$

where n is an integer. This formula is confirmed by wave mechanics. It is in agreement with experiment, not only for the spectra of atomic hydrogen $(Z = 1)$ and ionised helium $(Z = 2)$, but for the X-ray levels of heavy atoms. For these it is a fair approximation to treat each K electron as moving in a field of a point charge $(Z-\sigma)e$, with σ between 0 and 1.

(b) That if one takes into account the motion of the nucleus (mass M) about the centre of gravity, m in the above equation must be replaced by m^*, where

$$m^* = m\Big/\left(1+\frac{m}{M}\right).$$

This small correction can be verified by comparing the values of the Rydberg constant obtained from hydrogen and ionised helium (cf. Chap. V, § 2·1).

(c) That a diatomic molecule rotating about its centre of gravity has quantised energy levels obtained as follows. We may treat it as a rigid body of moment of inertia I, given by

$$I = 2Ma^2;$$

$2a$ is here the distance between the nuclei and M the mass of each nucleus, the electronic mass being neglected. If the

† Cf. the original papers by Niels Bohr, or any text-book on atomic physics.

angular velocity about the centre of gravity is ω, the angular momentum is $I\omega$, so that Bohr's hypothesis gives us

$$I\omega = l\hbar,$$

where l is an integer. The kinetic energy W is thus given by

$$W = \tfrac{1}{2}I\omega^2 = \frac{l^2\hbar^2}{2I}. \tag{1}$$

The treatment by the methods of wave mechanics replaces l^2 by $l(l+1)$, as shown in Chapter V, § 2·2.

It will be noticed that the interval between energy levels is less by a factor of order m/M (1/1860 for hydrogen) than for the line spectra of free atoms. For if we take for a the radius of the first Bohr orbit of hydrogen ($\hbar^2/me^2$) we find for the interval ΔW between the ground state ($l = 0$) and the first first excited state ($l = 1$)

$$\Delta W = \frac{\hbar^2}{2I} = \frac{m}{2M}\frac{me^4}{2\hbar^2}.$$

The second factor is the ionisation energy of hydrogen (13·60 eV.).

ΔW is thus less by a considerable factor than kT at room temperature (0·025 eV.). Therefore diatomic molecules in a gas in thermal equilibrium at room temperature will rotate, as is shown by the observed value $\tfrac{5}{2}R$ of the specific heat. At very low temperatures the specific heat of H_2 does in fact drop towards the value $\tfrac{3}{2}R$, since when $kT < W$ the collisions are not in general energetic enough to excite rotation.† Or, expressed in other terms, the exponential factor $\exp(-W/kT)$, which determines the number of rotating molecules, becomes small.

2. TREATMENT OF STATIONARY STATES IN WAVE MECHANICS

According to the principles of wave mechanics, any electron or other particle shut up in an enclosed space will have a

† The experimental evidence is reviewed, for instance, by R. H. Fowler, *Statistical Mechanics*, 2nd ed., Cambridge, 1936, p. 83.

quantised series of energy values. Thus an electron bound in an atom, or an atom vibrating as a whole about a fixed position in a solid, will have quantised energy values, while a freely moving atom or electron will not.

The reason for this can be seen most simply by considering an idealised case, that of the motion of a particle shut up in a box with perfectly reflecting sides. We need to consider motion in one dimension only; we thus consider that the particle moves along the x-axis from $x = 0$ to $x = a$, and that at these two extremities, where the walls of the box are, the wave function vanishes. The Schrödinger equation for the particle is

$$\frac{d^2\psi}{dx^2} + \frac{2mW}{\hbar^2}\psi = 0,$$

of which the solutions are

$$\sin kx, \quad \cos kx,$$

where
$$k^2 = \frac{2mW}{\hbar^2}.$$

The solution which vanishes at $x = 0$ is $\sin kx$, and this vanishes at $x = a$ only if $ka = n\pi$, and thus if

$$W = \frac{\hbar^2 n^2 \pi^2}{2ma^2} = \frac{h^2 n^2}{8ma^2}. \tag{2}$$

Solutions of the Schrödinger equation satisfying the required boundary conditions exist therefore only if W has the series of quantised values given by (2).

It will be remembered that, according to wave mechanics, our knowledge of the position of a particle in a given state is given by a wave function ψ, such that $|\psi|^2 dx$ is the probability that the particle will be found between x and $x + dx$. In our case

$$\psi = C \sin \frac{n\pi x}{a},$$

where C is a constant. Such a solution can only be found if W has one of the values (2). We deduce that the energy *can* only have these values.

The value of C should be chosen so that the function is normalised, i.e. so that

$$\int_0^a \psi^2 dx = 1.$$

As regards orders of magnitude, we note that if a is of the dimensions of an atom (3×10^{-8} cm.), and m the mass of an electron, the quantity $h^2/8ma^2$ is of the order of the ionisation potential of an atom (actually 4·1 eV.). If a is the diameter of a nucleus (say 5×10^{-13} cm.), and m the mass of a nucleon, we obtain c. 10^7 eV., of the order of the energies concerned in nuclear reactions.

Suppose that we now consider an electron shut up in a box bounded, not by perfectly reflecting sides, but by an electrostatic field which pushes the electron back when it tries to get out. The potential energy of an electron in this field is shown in Fig. 15. Let

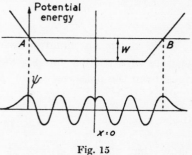

Fig. 15

the electron have an arbitrary energy W. The wave function in the neighbourhood of the points A and B, where the 'classical' electron would try to get out, is shown also in Fig. 15; in the regions where the classical electron cannot go, ψ will decrease exponentially, while within the box it will oscillate. If we start to draw the wave function from either end taking the solutions that decay exponentially outside the box instead of increasing exponentially, the two solutions will not in general join up in the middle; only for a discrete series of energy values W_n will they do so, and these will be the quantised values that the energy of the electron must have.

An electron in a hydrogen atom is held in a box very much of this type. The potential energy of such an electron plotted along a line passing through the nucleus is shown in the upper

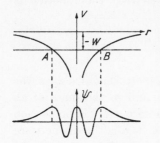

part of Fig. 16: an electron with the energy represented by the horizontal line AB can move freely between the points A, B, where it will suffer total reflection. The wave function will be as shown in the lower half of the diagram; it will oscillate in the region between A and B and die away exponentially outside. Clearly, only for a series of energies W_n will such a solution of the Schrödinger equation be obtainable.

Fig. 16. Potential energy of electron and wave function for hydrogen atom.

Exercise

Write down an equation whose roots are the quantised energies of a particle moving along the x-axis in the field derived from the potential

$$V(x) = 0 \qquad |x| > \alpha,$$
$$= -U \qquad |x| < \alpha.$$

Show that there is always at least one bound state for a particle in this field, but only one if

$$\alpha \sqrt{(2mU)}/\hbar < \pi. \tag{3}$$

3. THE SIMPLE HARMONIC OSCILLATOR

One of the simplest and most important examples in the theory of stationary states is that of the linear oscillator. A particle of mass M is held to a fixed point P by a restoring force $-px$ where x is the displacement.† According to

---

† The most important applications are the vibrations of an atom in a solid, where x is the displacement, and the vibrations of a diatomic molecule, where x is the change in the internuclear distance.

Newtonian mechanics, it will vibrate about P with arbitrary amplitude and energy, and frequency ν given by

$$\nu = \frac{1}{2\pi} \sqrt{\frac{p}{M}}.$$

Our problem in quantum mechanics is to find the allowed quantised values of the energy W, and the appropriate wave functions.

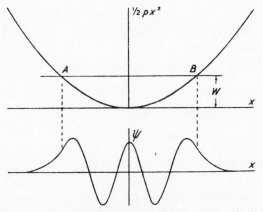

Fig. 17. Potential energy of a particle executing simple harmonic motion.

The potential energy of the particle is $\frac{1}{2}px^2$. This is plotted in Fig. 17; according to classical mechanics the particle will move backwards and forwards between A and B. Since the velocity of the particle is $\sqrt{\{2(W-V)m\}}$, the chance that it will be found between x and $x+dx$ is proportional to $(W-V)^{-\frac{1}{2}}$.

The Schrödinger equation is

$$\frac{d^2\psi}{dx^2} + \frac{2M}{\hbar^2}\left(W - \tfrac{1}{2}px^2\right)\psi = 0. \tag{4}$$

It is interesting to obtain an approximate solution by means of the W.K.B. method. This is (for x between A and B)

$$(W-V)^{-\frac{1}{4}}\sin\int^x \sqrt{\left\{\frac{2M}{\hbar^2}\left(W - \tfrac{1}{2}px^2\right)\right\}}\,dx. \tag{5}$$

The true solution will decrease exponentially outside AB. As an approximation, however, we may demand that the wave function shall vanish at the two extremities A, B. The lower limit of the integral in (5) must then be taken at A, and the condition is thus satisfied if

$$\int_{-x_1}^{x_1} \sqrt{\left\{\frac{2M}{\hbar^2}\left(W - \tfrac{1}{2}px^2\right)\right\}} \, dx = n\pi,$$

$\pm x_1$ being the values of x for which the integrand vanishes. Our quantum condition, then, consists in fitting n half waves into the interval AB.

The integral can be evaluated by setting

$$\tfrac{1}{2}px^2 = W \sin^2 \theta;$$

we find, as may easily be seen,

$$W = nh\nu, \quad n \geqslant 1.$$

The exact solution of the problem involves finding the values of W for which a solution of (4) exists which oscillates between A and B and decays exponentially outside. These values of W are given by

$$W = (n + \tfrac{1}{2}) h\nu.$$

The normalised solution† for the first two states are

$$\pi^{-\frac{1}{4}} \exp\left(-\tfrac{1}{2}y^2\right),$$

$$(4/\pi)^{\frac{1}{4}} y \exp\left(-\tfrac{1}{2}y^2\right),$$

where

$$y = \left(\frac{Mp}{\hbar^2}\right)^{-\frac{1}{4}} x.$$

The reader will easily verify that the above solutions satisfy (4). The quantity $(\hbar^2/Mp)^{\frac{1}{4}}$ gives a measure of the radial extension of the wave function in the ground state.

The quantum number n ($n = 1, 2, \ldots$) which labels each stationary state has in wave mechanics a simple meaning; $n-1$

† For details of the solution cf., for example, Mott and Sneddon, p. 51.

is equal to the number of zeros in the wave function, not counting the zeros at the two extremities.

It is worth emphasising here that, if we are dealing with an atom vibrating in a solid or molecule, the radial extension of the wave function in the ground state is small compared with the distance between atoms. For if this latter distance is a, we may expect the interatomic forces to be of order e^2/a^2, and thus

$$p \sim e^2/a^3.$$

The radial extension of the wave function, as we have seen, is $(\hbar^2/Mp)^{\frac{1}{4}}$; substituting for p this gives

$$a(\hbar^2/Me^2a)^{\frac{1}{4}}.$$

But $\hbar^2/me^2$ is the radius of the hydrogen atom, and thus of the same order as a. Thus the radial extension is of order

$$a(m/M)^{\frac{1}{4}},$$

where m is the mass of an electron, M of an atom.

4. QUANTISATION IN THREE DIMENSIONS

Quantisation in three dimensions differs from that in one in two respects: each stationary state is specified by three quantum numbers instead of one; and levels may be degenerate, that is to say two different states with different wave functions may have the same value of the energy.

These two points are both illustrated by the case of a particle moving in a box with perfectly reflecting sides. Let the interior of the box be defined by

$$0 < x < a,$$
$$0 < y < b,$$
$$0 < z < c,$$

so that the box has rectangular sides with edges of length a, b, c. The Schrödinger equation is

$$\nabla^2\psi + \frac{2mW}{\hbar^2}\,\psi = 0,$$

which has solutions satisfying the boundary conditions

$$\psi = A \sin\left(\frac{\pi n_1 x}{a}\right) \sin\left(\frac{\pi n_2 y}{b}\right) \sin\left(\frac{\pi n_3 z}{c}\right),$$

if, and only if, the energy W has one of the values

$$W = \frac{\pi^2 \hbar^2}{2m}\left(\frac{n_1^2}{a^2} + \frac{n_2^2}{b^2} + \frac{n_3^2}{c^2}\right). \tag{6}$$

The quantum numbers n_1, n_2, n_3, with unity subtracted in each case, are equal to the number of nodal planes in the wave function parallel to the planes $x = 0$,

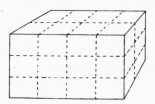

Fig. 18. Nodal planes of wave function of particle in a box.

$y = 0$, and $z = 0$ respectively. The case illustrated in Fig. 18 shows the state (4, 3, 2).

That degenerate states may occur is clear from formula (6). For instance, if $a = b$, the wave functions with quantum numbers (n_1, n_2, n_3) and (n_2, n_1, n_3) are solutions of the wave equation corresponding to the same value of the energy. It will be noticed that, when two degenerate states exist with wave functions ψ_1, ψ_2, then a linear combination such as

$$A\psi_1 + B\psi_2 \tag{7}$$

is also a solution of the wave equation.

Exercise

If $a = b$, sketch the nodal surfaces of the wave function (7) for $n_1 = 1$, $n_2 = 2$ and various values of the constants A, B.

5. QUANTISATION WITH SPHERICAL SYMMETRY AND THE HYDROGEN ATOM

This is an important problem, as it includes the treatment of the hydrogen and other atoms. We have to find the stationary states of a particle moving in three dimensions which is under the attraction of a centre of force. The potential energy of the particle is then a function only of its distance r from the centre of force. We denote it by $V(r)$. In the particular case of an electron moving in the field of a positively charged nucleus of charge Ze,

$$V(r) = \frac{-Ze^2}{r}. \tag{8}$$

The Schrödinger equation is then

$$\nabla^2 \psi + \frac{2m}{\hbar^2} \{W - V(r)\} \psi = 0.$$

The solutions of this equation in spherical harmonics can be found in numerous text-books and will not be given here. The salient points are given in the remainder of this section.

The solutions may be divided into:

(a) Solutions having spherical symmetry, such that

$$\psi = f(r).$$

The corresponding states of the atom are known as s states,† the quantum number l, giving the number of nodal surfaces passing through the origin, being zero. $l\hbar$ will later be identified with the angular momentum of the state.

(b) Solutions having the forms

$$\psi = \frac{x}{r} f(r), \quad \frac{y}{r} f(r), \quad \frac{z}{r} f(r).$$

These are known as p states; they have one nodal plane passing through the origin; thus by definition $l = 1$. A p state is triply degenerate, the three independent wave functions shown above all having the same energy.

(c) Solutions having the forms

$$\psi = \frac{yz}{r^2} f(r), \quad \frac{zx}{r^2} f(r), \quad \frac{xy}{r^2} f(r),$$

$$\frac{(y^2 - z^2)}{r^2} f(r), \quad \frac{(z^2 - x^2)}{r^2} f(r), \quad \frac{(x^2 - y^2)}{r^2} f(r).$$

† The symbols used in spectroscopy, s, p, d, referred originally to *lines*, not states, and meant sharp, principal, diffuse.

These are known as d states; they have two nodal planes passing through the origin, so that $l = 2$. The degeneracy is fivefold, not sixfold as would at first appear, because the three last wave functions are not independent, their sum being zero.

(d) States of higher quantum number l; these can conveniently be expressed only in terms of spherical harmonics

$$\psi = P_l^{|u|}(\cos \theta)\, e^{iu\phi} f(r) \quad -l \leqslant u \leqslant l.$$

The degree of degeneracy is $2l + 1$.

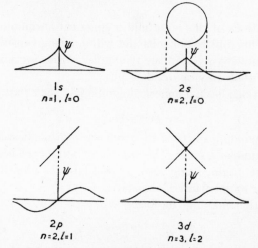

Fig. 19. Nodal surfaces and amplitude of the wave functions of the hydrogen atom.

The function $f(r)$ will itself have a number of zeros; each of these determines a spherical nodal surface in the wave function.

The principal quantum number, n, is defined so that $n-1$ is equal to the total number of nodal surfaces, planar and spherical. Thus $n-l-1$ is equal to the number of zeros in $f(r)$.

If $V(r)$ is given by (8) and thus is the potential energy in a Coulomb field, the energy of an electron depends only on n (apart from relativistic corrections mentioned in Chapter VII), and is given by

$$W = -\frac{me^4 Z^2}{2\hbar^2}\frac{1}{n^2}.$$

This is not the case for any other field, the energy depending on l also.

In Fig. 19 we show the wave functions of the hydrogen atom for a number of states. Above we show the intersection of the nodal surfaces with the plane passing through the nucleus.

Turning now to the Schrödinger equation, it will easily be seen by direct substitution that for s states, where $\psi = f(r)$, the equation reduces to

$$\frac{d^2f}{dr^2} + \frac{2}{r}\frac{df}{dr} + \frac{2m}{\hbar^2}(W-V)f = 0.$$

The student is recommended to verify this. In the general case it becomes

$$\frac{d^2f}{dr^2} + \frac{2}{r}\frac{df}{dr} + \left\{\frac{2m}{\hbar^2}(W-V) - \frac{l(l+1)}{r^2}\right\}f = 0. \tag{9}$$

The substitution $\qquad f = \dfrac{g}{r}$

reduces (9) to

$$\frac{d^2g}{dr^2} + \left\{\frac{2m}{\hbar^2}(W-V) - \frac{l(l+1)}{r^2}\right\}g = 0. \tag{10}$$

This is of the same form as the equation for the motion of an electron in one dimension. The solution g, however, must vanish at the origin, in order that f may remain finite there. The type of solution is shown in Fig. 20. The quantity $F(r)$ defined by

$$F(r) = \frac{2m}{\hbar^2}(W-V) - \frac{l(l+1)}{r^2}$$

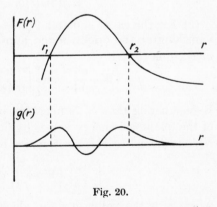

Fig. 20.

is positive between two values r_1 and r_2; within this region the wave function oscillates; outside it decays exponentially.

For the case of the Coulomb field, equation (9) may be solved in series. The solution may be found in any text-book; we quote here the first three wave functions

$$1s \quad (n = 1, l = 0) \quad f(r) = c_1 e^{-r/a},$$

$$2s \quad (n = 2, l = 0) \quad f(r) = c_2 (2 - \frac{r}{a}) e^{-r/2a},$$

$$2p \quad (n = 2, l = 1) \quad f(r) = c_3 r e^{-r/2a}.$$

Here a is the Bohr radius, given by

$$a = \hbar^2 / m Z e^2,$$

and the constants c_1, c_2, c_3 are normalising factors. The student is recommended to verify by substitution that these values of $f(r)$ are solutions of equation (9) for the appropriate values of W.

Exercises

(1) Consider the motion of an electron in the field defined by

$$V(r) = 0 \qquad r > \alpha,$$
$$= -U \quad r < \alpha.$$

Show that there are no stationary states at all if

$$\alpha \sqrt{(2mU)}/\hbar < \tfrac{1}{2}\pi.$$

(2) For the case $l = 0$ and $V(r) = -e^2/r$, the solution of (10), vanishing at the origin, found approximately by the W.K.B. method, is

$$g = \left\{ W + \frac{e^2}{r} \right\}^{-\frac{1}{4}} \sin \left[\int_0^r \left\{ \frac{2m}{\hbar^2} \left(W + \frac{e^2}{r} \right) \right\}^{\frac{1}{2}} dr \right].$$

Remembering that W is negative, obtain approximate values of the quantised energy values by fitting n half-waves into the region where $W - V$ is positive, i.e. by setting

$$\int_0^{r_0} \left\{ \frac{2m}{\hbar^2} \left(W + \frac{e^2}{r} \right) \right\}^{\frac{1}{2}} dr = n\pi,$$

where r_0 is the value of r for which the integrand vanishes.†

† The method in fact gives the exact value of W. This would not be the case for any other form of $V(r)$.

The integral can be evaluated by setting

$$\frac{e^2}{r} + W = x^2.$$

(3) Show that if $V(r) \rightarrow -\text{const.}/r$ at large values of r, there will exist an infinite series of stationary states leading up to a series limit; while if $V(r)$ tends to zero faster than r^{-1}, the number of bound states is finite.

(4) Show that, for an attractive field of the type

$$V(r) = -a/r^2,$$

the equation giving the energy values is indeterminate.

6. INTERPRETATION OF THE WAVE FUNCTIONS

If a hydrogen atom is, for example, in the ground state, it is described by a wave function ψ given by

$$\psi(r) = Ce^{-r/a}, \quad a = \hbar^2/me^2 = 0 \cdot 54A.$$

The interpretation of this function is, as usual, that $|\psi(r)|^2 d\tau$ is the probability that an electron will be found in the volume element $d\tau$ at a distance r from the nucleus. This is all the information provided by wave mechanics about the position of the electron.

It is usual to choose the constant C so that

$$\int |\psi(r)|^2 d\tau = 1; \tag{11}$$

the wave function will then describe a state of affairs where it is known that an electron is in the atom. The wave function is then said to be normalised. It will easily be verified that

$$C = \pi^{-\frac{1}{2}} a^{-\frac{3}{2}}.$$

In some phenomena (e.g. scattering problems, cf. Chap. II, § 7) the atom behaves like a nucleus surrounded by a cloud of negative charge of density

$$-e |\psi(r)|^2,$$

where ψ is the normalised wave function. The electrostatic potential of the field due to a nucleus and this charge is, as may easily be verified,

$$e\left(\frac{1}{r}+\frac{1}{a}\right)e^{-2r/a}.$$

A second electron in the neighbourhood of a hydrogen atom may thus be treated as though it moved in a field of potential energy $V(r)$ such that

$$V(r) = -e^2\left(\frac{1}{r}+\frac{1}{a}\right)e^{-2r/a}.$$

In atoms containing more than one electron it is a fair approximation to consider each electron as having its own wave function $\psi(r)$. Thus, for instance, in the helium atom each electron may be regarded as having a wave function

$$\psi(r) = \left(\frac{1}{\pi a^3}\right)^{\frac{1}{2}}e^{-r/a}, \tag{12}$$

where a is a parameter. One way of finding the best value of the parameter a is the variational method given in the next section.

Another method of finding the wave functions $\psi(r)$ is that due to Hartree, the method of the 'self-consistent field'. Hartree does not use an analytic form for $\psi(r)$; for the helium atom he proceeds as follows. He supposes that each electron moves in a field of potential energy $V(r)$ produced by the nucleus (charge $Z = 2$) and by the other electron, treated as a distribution of charge of density $e\,|\,\psi(r)\,|^2$. $V(r)$ will then be given by Poisson's equation

$$\frac{d^2V}{dr^2}+\frac{2}{r}\frac{dV}{dr} = 4\pi\,e^2|\,\psi(r)\,|^2$$

and the boundary conditions

$$V(r) \sim -2e^2/r \quad r \text{ small,}$$
$$\sim -e^2/r \quad\ r \text{ large.}$$

If we know ψ, then, we can calculate $V(r)$; and if we know $V(r)$ we can calculate ψ from Schrödinger's equation. Hartree, by a process of successive approximations, calculates ψ so that it is self-consistent; in other words, if V is deduced from ψ and then ψ calculated, the original value is obtained.†

7. VARIATIONAL METHOD OF OBTAINING APPROXIMATE WAVE FUNCTIONS

Let us write the Schrödinger equation for a single particle in a field with potential energy $V(r)$ in the form (cf. p. 46)

$$(H - W)\psi = 0,$$

where

$$H = -\frac{\hbar^2}{2m} \nabla^2 + V.$$

Multiplying by ψ^* and integrating over all space we see that

$$W = \int \psi^* H \psi \, d\tau. \tag{13}$$

The total energy of the electron is thus the sum of two parts:

(a) $\int \psi^* V \psi \, d\tau$. This clearly represents the potential energy of the electron in the field of potential energy $V(r)$.

(b) $-\frac{\hbar^2}{2m} \int \psi^* \nabla^2 \psi \, d\tau$. This (positive) term represents the kinetic energy. The reasons why this is so lie somewhat outside the scope of this book.

It is shown in the more advanced books on wave mechanics that if $\psi(x, y, z)$ is *any* normalised function of x, y, z, i.e. such that (11) is satisfied, then the true wave function is that for which the integral (13) is a minimum. This gives a very useful method of obtaining approximate wave functions; one can take a given analytic expression containing one or more parameters, work out the integral (13) and choose the parameters so as to minimise the energy.

† Hartree's method is discussed further in Chap. V; a review is given by D. R. Hartree, *Rep. Prog. Phys.* XI, 113, 1946-7.

Exercises

(1) For the hydrogen atom, take ψ of the form $Ce^{-\lambda r}$ normalised to unity and work out the integral (13). Show that it has the minimum value when $\lambda = me^2/\hbar^2$.

(2) A hydrogen atom is placed in a weak electric field E, so that the potential energy of its electron is

$$-\frac{e^2}{r} + Eex.$$

By assuming a wave function of the form

$$(A + Bx)\, e^{-\lambda r},$$

obtain an approximate expression for the energy due to the field.

(3) Describing both electrons in the helium atom by a wave function ψ of the form (12) with an unknown value of a, calculate the first ionisation energy. This is done as follows. First calculate the kinetic energy and potential energy in the field of the nucleus. Their sum is

$$2\int\psi^*\left(-\frac{\hbar^2}{2m}\nabla^2 - \frac{2e^2}{r}\right)\psi\,d\tau,$$

the 2 arising because there are two electrons. Then calculate the interaction energy of the electrons; this, by the arguments of the last section, is

$$\int\psi^*\left\{-e^2\left(\frac{1}{r} + \frac{1}{a}\right)e^{-2r/a} + \frac{e^2}{r}\right\}\psi\,d\tau.$$

The sum of these is the total (negative) energy of the atom, and a must be chosen so that it takes a minimum value. To obtain the first ionisation energy, we must subtract it from the second ionisation energy, viz. $4 \times me^4/2\hbar^2$.

8. ORTHOGONAL PROPERTY OF THE WAVE FUNCTIONS

This property is of considerable importance for subsequent developments. The term orthogonal means that, if ψ_1, ψ_2 are solutions of the Schrödinger equation corresponding to two different energy values, then

$$\int \psi_1^* \psi_2 d\tau = 0,$$

the integral being over all space.

Exercise

The student is recommended to verify the orthogonal relation for the functions shown on pp. 60, 66. It is, for instance, obvious from symmetry that s and p functions are orthogonal to each other.

We shall prove the orthogonal relation for the case of one dimension only, that is to say for a particle moving along a straight line. The two wave functions then satisfy the equations

$$\frac{d^2\psi_1}{dx^2} + \frac{2m}{\hbar^2}(W_1 - V)\psi_1 = 0,$$

$$\frac{d^2\psi_2^*}{dx^2} + \frac{2m}{\hbar^2}(W_2 - V)\psi_2^* = 0.$$

If we multiply the first equation by ψ_2^* and the second by ψ_1 and subtract the second from the first, we obtain

$$\psi_2^* \frac{d^2\psi_1}{dx^2} - \psi_1 \frac{d^2\psi_2^*}{dx^2} + \frac{2m}{\hbar^2}(W_1 - W_2)\psi_2^* \psi_1 = 0.$$

The first two terms may be written

$$\frac{d}{dx}\left(\psi_2^* \frac{d\psi_1}{dx} - \psi_1 \frac{d\psi_2^*}{dx}\right).$$

Thus on integrating over all space

$$\left[\psi_2^* \frac{d\psi_1}{dx} - \psi_1 \frac{d\psi_2^*}{dx}\right]_{-\infty}^{\infty} = -\frac{2m}{\hbar^2}(W_1 - W_2)\int \psi_2^* \psi_1 dx.$$

The term on the left vanishes, because, since we are dealing with bound electrons, the wave functions tend to zero far from the region in which they are bound. Also we have postulated that W_1 and W_2 should be unequal. It follows that

$$\int_{-\infty}^{\infty} \psi_2^* \psi_1 \, dx = 0. \qquad (14)$$

We have taken the complex conjugate of ψ_2 rather than the function itself. For a particle moving in an electrostatic field the wave functions are real, and this makes no difference. In the presence of a magnetic field, however, terms of the type $Ai\,\partial\psi/\partial x$ occur in the wave equation (Chap. II, §8), so the wave functions are complex. The reader will easily verify from Chapter II (13) that if H is a constant the relation (14) is verified in this case.

An important consequence of the orthogonal relation is that any arbitrary function $f(x, y, z)$ can be expanded in a series of wave functions ψ_n, provided that the function f satisfies the same boundary conditions as the functions ψ_n. Thus suppose, for instance, that the functions $\psi_n(x, y, z)$ are the series of wave functions for the hydrogen atom, and that $f(x, y, z)$ is any function which tends to zero at infinity. Then we may write

$$f(x, y, z) = \sum_n A_n \psi_n(x, y, z), \qquad (15)$$

and find the coefficients A_n by multiplying both sides by ψ_m^* and integrating over all space. We find, using the orthogonal relation, that

$$A_m = \int f(x, y, z) \, \psi_m^*(x, y, z) \, d\tau.$$

For a particle bound in a box, a simple harmonic oscillator, etc., all the states are quantised. For a hydrogen atom, on the other hand, the summation should include an integration over the unquantised states, in which the electron is free from the atom (cf. Chap. VI, §3).

It will be shown in Chapter VI that if an electron is described by a wave function of the type (15), the coefficients A_n may be interpreted as follows: we do not know in which stationary state the electron is; the chance that it is in the nth is $|A_n|^2$.

9. PERTURBATION THEORY

It is very useful to be able to calculate the change in the energy of an electron in an atom due to a small perturbation, for example, an electric or magnetic field. This may be done as follows:

We write the wave equation

$$(H - W)\psi = 0, \tag{16}$$

and calculate the change w in W due to a small additional term $v(x, y, z)$ added to the potential energy. The wave equation may now be written

$$\{H + v - (W + w)\}(\psi + f) = 0, \tag{17}$$

where f is the change in ψ. Making use of (16) and neglecting all terms of the second order such as vf, wf, this gives

$$(H - W)f + (v - w)\psi = 0.$$

Now we may expand f in a series of the type (15)

$$f = \sum_n A_n \psi_n, \tag{18}$$

where the ψ_n are the solutions of (16) and W_n the corresponding energy values; we use also the suffix zero to denote the state the electron is in.

We then find

$$\sum_n A_n (W_n - W_0) \psi_n + (v - w)\psi_0 = 0.$$

If we multiply both sides by ψ_0^* and integrate, we find

$$w = \int \psi_0^* v \psi_0 d\tau, \tag{19}$$

a formula which is essentially the same as (13), and gives us to the first order the change w in the energy. If we multiply both sides by ψ_m^* and integrate, we obtain

$$A_m = -\frac{\int \psi_m^* v \psi_0 d\tau}{W_m - W_0}, \tag{20}$$

which, with (18), gives us the perturbed wave function.

Exercises

(1) Let us suppose that the potential energy of an electron in the field of a nucleus is, owing to the finite radius r_0 of the nucleus,

$$-Ze^2/r \qquad r > r_0,$$
$$-Ze^2/r_0 \qquad r < r_0.$$

Calculate the change in the energy of the 1s state due to this correction.

(2) A particle is bound to its mean position by a force such that its potential energy is $\frac{1}{2}px^2 + qx^4$. Treating the second term as a perturbation, find the energies of the first two stationary states.

Two of the most important applications of perturbation theory are to an electron in an electric and in a magnetic field; these will be treated in the next three sections.

10. THE POLARISABILITY OF AN ATOM

In this section we shall calculate the polarisability of an atom. The calculation will be specifically for a hydrogen atom, but can be extended to any atom if we describe each electron by its own wave function (p. 102).

We give first a classical calculation. Suppose that an electron is bound to a mean position by a force $-px$ when the displacement is x. It can then vibrate with a frequency ν_0 given by

$$\nu_0 = \frac{1}{2\pi}\sqrt{\frac{p}{m}}.$$

In the presence of a field E, the displacement x is given by

$$px = eE. \tag{21}$$

The dipole moment is thus

$$ex = e^2 E/p = e^2 E/4\pi^2 m\nu_0^2.$$

The polarisability α, defined as ex/E, is thus given by

$$\alpha = \frac{e^2}{4\pi^2 m\nu_0^2}. \tag{22}$$

If, instead of a static field, E is replaced by a field $E_0 \cos 2\pi\nu t$ oscillating with frequency ν, (21) becomes

$$m\ddot{x} + px = eE_0 \cos 2\pi\nu t;$$

on integrating we find

$$ex = e^2 E_0 \cos 2\pi\nu t / 4\pi^2 m(\nu_0^2 - \nu^2),$$

so that the polarisability is now given by

$$\alpha = \frac{e^2}{4\pi^2 m(\nu_0^2 - \nu^2)}. \tag{23}$$

This calculation is artificial in that it makes use of the concept of an elastically bound electron, capable of vibrating with a definite frequency. We shall now give the corresponding wave-mechanical calculation.

The electron in the atom is acted on by a field E in, for instance, the z direction. The additional term in the potential energy of the electron, due to this field, is†

$$v(z) = Eez.$$

In the presence of the field, then, the wave function of the electron is, instead of the original $\psi_0(r)$, by (18) and (20)

$$\psi_0(r) - eE \sum_n \frac{z_{n0}}{W_n - W_0} \psi_n(x, y, z), \tag{24}$$

where z_{n0} is defined by

$$z_{n0} = \int \psi_n^* z \psi_0 \, d\tau,$$

and $d\tau$ denotes the element of volume $dx\,dy\,dz$. The charge density $\rho(x, y, z)$ in the atom is obtained by taking the square of the modulus of (24) and multiplying by $-e$, so that

$$\rho(x, y, z) = -e\,|\psi_0(r)|^2 + e^2 E \sum_n \frac{z_{0n}}{W_n - W_0} \{\psi_n^* \psi_0 + \psi_n \psi_0^*\} \tag{25}$$

Terms of order E^2 are neglected, since the polarisability is always defined for values of E small compared with the fields within the atom.

† Note that e is the numerical value of the electronic charge, $-e$ the charge on the electron.

Now the dipole moment of the atom is by definition

$$\int \rho(x, y, z)\, z\, d\tau, \tag{26}$$

so that the polarisability α is $\int \rho z\, d\tau / E$. The function $|\psi_0(r)|^2$ is spherically symmetrical; thus, substituting (25) in (26), the first term gives a vanishing contribution, and we are left with

$$\alpha = 2e^2 \sum_n \frac{|z_{n0}|^2}{W_n - W_0}, \tag{27}$$

which gives us the desired quantum mechanical formula for the polarisability of an atom.

This formula (27) may be compared with the classical formula (22). The frequencies ν_{n0} of the absorption lines of the atom are given by

$$h\nu_{n0} = W_n - W_0.$$

Thus (27) may be written

$$\alpha = \frac{e^2}{4\pi^2 m} \sum_n \frac{f_{n0}}{\nu_{n0}^2},$$

where

$$f_{n0} = \frac{8\pi^2 m \nu_{n0}}{h} |z_{n0}|^2.$$

It may also be shown, though the calculation will not be given here, that for a field vibrating with frequency ν, the formula for the polarisability which generalises (23) is

$$\alpha = \frac{e^2}{4\pi^2 m} \sum_n \frac{f_{n0}}{\nu_{n0}^2 - \nu^2}. \tag{28}$$

The quantity f_{n0} is called the 'oscillator strength' of the transition. It may easily be proved† that, for atoms containing a single electron,

$$\sum_n f_{n0} = 1. \tag{29}$$

That this relation is satisfied follows also from (28), since for high values of the frequency ν the polarisability must tend to

† Cf., for example, Mott and Sneddon, p. 169.

the classical value (23), the forces binding the electron to the atom being then unimportant.

It will be shown in Chapter VI that the intensities of absorption and emission lines are proportional to the oscillator strengths.

It will be seen at once by symmetry that for the hydrogen atom in the ground state all oscillator strengths vanish except those for which n refers to a p state. The values of the non-vanishing oscillator strengths have been calculated.† Some values are, for transitions from the initial state $1s$:

Final state	f_{n0},
$n = 2$	0·4161,
$n = 3$	0·0791,
Asymptotic formula for large n	$1·6n^{-3}$,
Σf_n for all discrete values of n	0·5641,
Σf_n for continuous spectrum	0·4359.

The dielectric constant κ, equal to the square of the refractive index, μ^2, can be deduced from the value of α by the formula
$$\kappa = \mu^2 = 1 + 4\pi N\alpha,$$
where N is the number of atoms per unit volume. It will be seen that, if $\nu \gg \nu_{n0}$, this formula tends to
$$\mu^2 = 1 - \frac{Ne^2}{\pi m \nu^2},$$
the usual formula for a medium containing N free electrons per unit volume.

It is also of interest to calculate the amplitude of the radiation scattered by a single atom. In the presence of an oscillating field $\mathbf{E}$, the dipole induced in the atom is $\alpha\mathbf{E}$. The electric vector of the scattered radiation, at a point P at a distance r from the atom in a direction making an angle ϕ with $\mathbf{E}$, is
$$\frac{\alpha E}{r} \frac{4\pi^2 \nu^2}{c^2} \sin\phi.$$

† Cf., for instance, H. Bethe, *Handb. Phys.* xxiv, pt. 1, 443, 1933.

Inserting formula (28) for α we find for the electric vector of the scattered radiation

$$\frac{E}{r}\frac{e^2}{mc^2}\nu^2\sin\phi\sum_n\frac{f_{n0}}{\nu_{n0}^2-\nu^2}.$$

It will be noticed that, if the incident radiation is on the long wavelength side of all absorption lines, the scattering increases with decreasing wavelength.

If $\nu\gg\nu_{n0}$ for all absorption lines of appreciable oscillator strength, the formula becomes

$$\frac{E}{r}\frac{e^2}{mc^2}\sin\phi, \tag{30}$$

the scattering formula for a free electron. Such a formula is obviously only valid if the wavelength of the radiation remains great compared with the size of the atom. For the case when this is not so, compare Chapter V, § 4·6.

Exercise

Work out the oscillator strength for the transition $2p$ to $1s$ of the hydrogen atom, using the wave functions given on p. 66.

11. THE STARK EFFECT

The threefold degenerate p states are split by an electric field. This may be seen as follows. The energy of the atom due to the electric field is $\frac{1}{2}\alpha E^2$. If the atom is in a p state, the wave functions are of the form $xf(r)$, $yf(r)$, $zf(r)$. Thus, if the field E is along the z-axis, the oscillator strength of the transition $p\to 1s$, for instance, vanishes except for the last of these wave functions. Thus for a field in the z-direction, the polarisability of the state $zf(r)$ differs from that for the other two. The three p states therefore split into one non-degenerate and one doubly degenerate state.

Consequently $p\to s$ transitions appear as doublets.

12. EFFECT OF A MAGNETIC FIELD

In this section we determine the effect of a magnetic field on an atom in a p state, obtain an expression for the Zeeman effect for an electron without spin, introduce the electron spin and describe the spin doublet in X-ray and optical spectra.

Suppose a magnetic field H is applied along the z-axis. Then there are three ways in which we can treat it: (a) by classical theory, (b) by old quantum theory, (c) by wave mechanics.

(a) *By classical theory.* We suppose (as on p. 74) that the electron is held in position by an elastic force $-pr$ for a displacement r, so that it can vibrate with frequency $\nu_0 = (2\pi)^{-1}\sqrt{(p/m)}$. In the presence of the magnetic field there are three possible normal modes, with different frequencies. The electron can vibrate along the field, in which case the frequency is unaltered, or rotate in circular orbits in the plane perpendicular to the field. In the latter case, if ω is the angular velocity of the motion and r the radius, we have

$$m\omega^2 r = pr \pm eH\omega r/c.$$

The first term represents the centrifugal force, the last the force acting on an electron moving with velocity ωr perpendicular to a magnetic field. Solving for ω, and treating H as a small quantity, we find

$$\nu = \frac{\omega}{2\pi} = \nu_0 \pm \frac{eH}{4\pi mc}. \tag{31}$$

If emission or absorption lines were due to vibrating electrons, we should expect that a magnetic field would split all lines into three components. Such a splitting is observed in a magnetic field (Zeeman effect), but only for singlet lines is it given by (31). The historical importance of the Zeeman effect is that the occurrence of e/m in a formula in agreement with experiment in certain cases gave the first experimental proof that electrons take part in the emission of light from atoms.

(b) *By old quantum theory.* One can show that the magnetic moment of an atom in which an electron rotates with angular momentum I is $eI/2mc$. For a circular orbit the proof is simple.

The magnetic moment is the product of the area (πr^2) and the current in electromagnetic units, which, if e is as usual in electrostatic units, is equal to

$$\pi r^2 \times e\omega/2\pi c = e\omega r^2/2c = eI/2mc.$$

The magnetic moment of an atom with angular momentum $l\hbar$ is thus

$$\frac{e\hbar}{2mc} l.$$

The quantity $e\hbar/2mc$ is known as the Bohr magneton and will be denoted by μ_B.

If one supposes that the component $u\hbar$ of the angular momentum along the magnetic field is also quantised and u takes integral values, it follows that the energy of the atom in the magnetic field is

$$u\mu_B H \quad -l \leqslant u \leqslant l.$$

The level thus splits in $2l + 1$ states (cf. § 5).

Making use of the selection rule (Chap. VI, § 7) that in optical transitions u will change only by 0 or ± 1, we see that a spectral line of frequency ν_0 will, in the presence of a magnetic field, split into three lines; the frequency of one of these is unaltered, while those of the others differ from ν_0 by $\Delta\nu$ where

$$h\Delta\nu = \pm \mu_B H.$$

The predictions of the old quantum theory for an electron without spin are thus the same as (31).

(c) *By wave mechanics.* This again gives the same result, and identifies the angular momentum $l\hbar$ of the last section with the quantum number l of p. 64. We shall, however, confine ourselves to p states ($l = 1$).

We have expressed the three wave functions of the p state in the form

$$\frac{x}{r}f(r), \quad \frac{y}{r}f(r), \quad \frac{z}{r}f(r).$$

Taking spherical polar coordinates these can, however, be written

$$\sin\theta\cos\phi f(r), \quad \sin\theta\sin\phi f(r), \quad \cos\theta f(r).$$

ϕ is here the azimuthal angle *about* the magnetic field. If, then, we want to represent rotation of the electron about this field, it is clear that the correct combinations of these wave functions to represent the three normal modes as in (a) above are

$$\psi_1 = \sin \theta e^{i\phi} f(r), \quad \psi_2 = \sin \theta e^{-i\phi} f(r), \quad \psi_3 = \cos \theta f(r). \quad (32)$$

Now the perturbing term due to a magnetic field H along the z-axis is (cf. Chap. II, § 7)

$$U = i\hbar\omega \frac{\partial}{\partial \phi}, \quad \omega = \frac{eH}{2mc},$$

and the changes in the energy of these three states due to the field are thus

$$\int \psi_n^* U \psi_n d\tau \quad n = 1, 2, 3.$$

The wave functions being normalised, this gives for the three states of (32)

$$\pm H\mu_B, \quad 0,$$

the same result as on the old theory.

This treatment, moreover, identifies l with the angular momentum of the atom.

13. THE ELECTRONIC SPIN

The hypothesis that the electron possesses a mechanical moment (angular momentum) equal to one half quantum ($\frac{1}{2}\hbar$), and a magnetic moment μ_B (one Bohr magneton) has to be introduced into physics for the following reasons:

(a) According to wave mechanics atoms in which one electron is outside a closed shell (e.g. Na, Ag) should in their normal states have the quantum number l equal to zero and should thus have no magnetic moment. The experiments of Gerlach and Stern on the splitting of beams of atoms by an inhomogeneous magnetic field show, however, that for the silver atom, for example, the ground state splits into *two* in a magnetic field. Since the multiplicity of a state with angular momentum $l\hbar$ is $2l+1$, one has to assume that the angular momentum of the atom is $\frac{1}{2}\hbar$, and ascribe this to the electron itself.

(b) Spectroscopic evidence shows that states for which $l > 0$ in atoms with one electron outside a closed shell are doublets. This can only be ascribed to the electron spin, which can have *two* orientations in the internal magnetic field which results from the orbital movement of the electron.

(c) Measurements of the gyromagnetic effect of ferromagnetic materials enable a value to be obtained for the small change in angular momentum of a specimen that accompanies a change in magnetic moment. For iron and nickel the ratio of magnetic to mechanical moment is e/mc, not $e/2mc$ as would be the case if the magnetism were due to the *movement* of the electrons. This proves the existence of elementary magnets (the electrons) for which the ratio is e/mc. Since the mechanical moment is $\frac{1}{2}\hbar$, the magnetic moment is $e\hbar/2mc$, or one Bohr magneton.

In describing the spin, then, we need to introduce a variable σ_z which can only take *two* values, ± 1. $-\sigma_z \mu_B H$ is defined as the energy which the electron's spin will have, if a magnetic field H is set up along the z-axis. It amounts to the same thing to say that $\frac{1}{2}\hbar\sigma_z$ is the component of the mechanical moment of the spin along the z-axis. The state of the electron will be described by a wave function $\chi(\sigma_z)$; the interpretation of this wave function is as usual that $|\chi(\sigma_z)|^2$ gives the probability that, if a measurement were made to determine the energy of the spin in a field H along the z-axis, the result would be $-H\mu_B\sigma_z$ ($\sigma_z = \pm 1$). There are two stationary states for $\chi(\sigma_z)$; the first, $\chi_\alpha(\sigma_z)$, is defined by the equations

$$\chi_\alpha(1) = 1, \quad \chi_\alpha(-1) = 0,$$

and describes the state of the spin when the energy is known to be $+\mu_B H$. The second, $\chi_\beta(\sigma_z)$, is defined by

$$\chi_\beta(1) = 0, \quad \chi_\beta(-1) = 1,$$

and describes the state of the spin when the energy is known to be $-\mu_B H$.

In this book these wave functions will be used only in describing the two-electron problem (cf. Chap. V, § 3).

The complete description of an electron is by the product of the orbital wave function $\psi(\mathbf{r})$ and the spin wave function $\chi(\sigma_z)$; thus

$$| \psi(\mathbf{r}) \chi(\sigma_z) |^2 d\tau$$

gives the probability that the electron is in the volume element $d\tau$ at the point $\mathbf{r}$, and at the same time the spin moment along the z-axis is $\frac{1}{2}\sigma_z \hbar$ ($\sigma_z = \pm 1$). The orbital function $\psi(\mathbf{r})$ will be little affected by the spin unless the electron is moving with velocity comparable with that of light, which for electrons bound in atoms is only the case for the inner X-ray levels. That the effect is small may be seen most simply as follows. If an electron is moving with velocity v it produces a magnetic field of order

$$H = ev/cr^2.$$

The energy of the electron's magnetic moment $e\hbar/2mc$ in such a field is of order

$$\frac{e^2 \hbar}{mr^2} \frac{v}{c^2}.$$

But, if r is the radius of the atom, h/mr is of the order v, where v is the velocity of an electron in the atom. Thus the energy term due to the electron's spin is of order

$$\frac{e^2}{r} \frac{v^2}{c^2},$$

which is smaller by the factor v^2/c^2 than the energy e^2/r of the electron in the field of the rest of the atom.

If we substitute $v \sim e^2/\hbar$, the term v^2/c^2 is seen to be of order $(e^2/\hbar c)^2$. The quantity $e^2/\hbar c$ is known as the fine structure constant, and is equal approximately to $1/137$.

For the calculation of the interaction between spin and orbital moment and for the evaluation, for instance, of the splitting of p states, one would naturally use the more exact theory of the electron due to Dirac, and reviewed briefly in Chapter VII. The analysis given here and the wave functions $\chi(\sigma)$ are more convenient, however, for an elementary treatment of the two-body problem, as given in the next chapter.

CHAPTER V

THE MANY-BODY PROBLEM

1. THE WAVE EQUATION FOR TWO PARTICLES

The problem considered in the preceding chapters has been the motion of a single particle in a field of force. The state of a particle has been described by an 'orbital' wave function $\psi(x, y, z)$ which depends on the spatial coordinates x, y, z, and by a spin wave function $\chi(\sigma_z)$ depending on the component σ_z of the spin moment along the z-axis. In this way a discussion of the hydrogen atom has been given by treating the electron as moving in the field of a fixed proton; and a discussion of more complicated atoms has been given by treating each electron as moving in the field of the nucleus and the *averaged* field of all the other electrons, so that each electron is given its separate wave function. This method is, of course, an approximation; in this chapter, then, we shall develop the theory appropriate to several interacting particles.

Let us consider two particles of masses m_1, m_2 moving along a straight line, and having coordinates x_1, x_2. Suppose also that the potential energy of the system when the two particles are at the points x_1, x_2 is $V(x_1, x_2)$. Then according to classical mechanics the equations of motion are

$$m_1 \ddot{x}_1 = -\frac{\partial V}{\partial x_1}, \quad m_2 \ddot{x}_2 = -\frac{\partial V}{\partial x_2}.$$

Now if we make the transformation

$$m_1^{\frac{1}{2}} x_1 = \xi_1, \quad m_2^{\frac{1}{2}} x_2 = \xi_2,$$

these equations transform into

$$\ddot{\xi}_1 = -\frac{\partial V}{\partial \xi_1}, \quad \ddot{\xi}_2 = -\frac{\partial V}{\partial \xi_2},$$

and are thus the same equations as those of a single particle of unit mass moving in two dimensions with coordinates ξ_1, ξ_2. This suggests that in wave mechanics also the treatment of two particles each moving on a straight line should be the same as that of one particle moving on a surface. If this is so the two particles will be described by a wave function $\psi(x_1, x_2)$, of which the interpretation will be the following: $|\psi(x_1, x_2)|^2 dx_1 dx_2$ is the probability that at any moment one particle will be found with its coordinate between x_1 and $x_1 + dx_1$ and the other particle between x_2 and $x_2 + dx_2$. Also this wave function will satisfy the equation of a particle of unit mass moving in a plane, namely

$$\frac{\hbar^2}{2} \left(\frac{\partial^2}{\partial \xi_1^2} + \frac{\partial^2}{\partial \xi_2^2} \right) \psi + (W - V) \psi = 0,$$

or, transforming back to the coordinates x_1, x_2,

$$\frac{\hbar^2}{2m_1} \frac{\partial^2 \psi}{\partial x_1^2} + \frac{\hbar^2}{2m_2} \frac{\partial^2 \psi}{\partial x_2^2} + (W - V) \psi = 0. \tag{1}$$

The treatment can be extended to the problem of two particles moving in three dimensions. Their behaviour should be determined by a wave function $\psi(x_1, y_1, z_1; x_2, y_2, z_2)$ of the co-ordinates of both particles. The interpretation of the wave function is that, if

$$P = |\psi(x_1, y_1, z_1; x_2, y_2, z_2)|^2 d\tau_1 d\tau_2, \tag{2}$$

then P is the probability that one particle will be found in the volume element $d\tau_1$ at the point (x_1, y_1, z_1) and the other in volume element $d\tau_2$ at the point (x_2, y_2, z_2). The wave function satisfies the equation

$$\frac{\hbar^2}{2m_1} \nabla_1^2 \psi + \frac{\hbar^2}{2m_2} \nabla_2^2 \psi + (W - V) \psi = 0, \tag{3}$$

which is the Schrödinger equation for a pair of particles. V is the potential energy of the particles, both in one another's field and in any external field.

The wave function for a pair of particles is thus a function of six coordinates. It will be realised that the 'wave' represented by this function is not a wave in any medium with extension in space.

It will be seen that if the two particles do not interact, and if one is in a state defined by a wave function $\psi_a(x_1, y_1, z_1)$, and the other in a state defined by a wave function $\psi_b(x_2, y_2, z_2)$, the wave function for the pair of particles is the *product*

$$\psi(x_1, y_1, z_1; x_2, y_2, z_2) = \psi_a(x_1, y_1, z_1) \, \psi_b(x_2, y_2, z_2). \qquad (4)$$

This is consistent with the interpretation of the wave functions, and can also be deduced from the wave equation, as the reader will easily verify.

If the two particles are of the same type (two electrons or two protons), the form is more complicated than (4), (cf. § 3).

2. A PAIR OF PARTICLES IN ONE ANOTHER'S FIELD

Under this heading we include

(*a*) the hydrogen atom, when the motion of the nucleus is considered,

(*b*) the rotation of diatomic molecules; in an elementary treatment we may suppose that the effect of the electrons is to introduce a force holding the two nuclei at a certain distance from each other.

The potential energy function V in (3) is then a function $V(r)$ of the distance r between the nuclei. The wave equation (3) can be separated by writing†

$$(m_1 + m_2) X = m_1 x_1 + m_2 x_2,$$
$$(m_1 + m_2) Y = m_1 y_1 + m_2 y_2,$$
$$(m_1 + m_2) Z = m_1 z_1 + m_2 z_2,$$
$$x = x_1 - x_2, \quad y = y_1 - y_2, \quad z = z_1 - z_2.$$

(X, Y, Z) are then the coordinates of the centre of gravity of the two particles, (x, y, z) the coordinates of one particle relative

† For a detailed treatment of this transformation, cf. A. Sommerfeld, *Wave Mechanics*, London, 1930, p. 27.

to axes through the other. The equation transforms into

$$\frac{\hbar^2}{2M_G}\left(\frac{\partial^2}{\partial X^2}+\frac{\partial^2}{\partial Y^2}+\frac{\partial^2}{\partial Z^2}\right)\psi+\frac{\hbar^2}{2m^*}\left(\frac{\partial^2}{\partial x^2}+\frac{\partial^2}{\partial y^2}+\frac{\partial^2}{\partial z^2}\right)\psi+(W-V)\psi=0,$$

where $M_G = m_1+m_2$, $m^* = m_1 m_2/(m_1+m_2)$.

Solutions of this equation may be obtained having the form

$$\psi = f(X, Y, Z)\, g(x, y, z),$$

where f satisfies $\qquad \dfrac{\hbar^2}{2M_G}\nabla^2 f + W_1 f = 0 \qquad$ (5)

and g satisfies $\qquad \dfrac{\hbar^2}{2m^*}\nabla^2 g + (W_2 - V)g = 0 \qquad$ (6)

if $W_1 + W_2 = W$. Clearly f is the wave function of a free particle of mass M, and represents the movement of the atom or molecule as a whole; g describes the probable length and orientation of the line joining the particles. Equation (6) is the same as that of a particle moving in a field in which its potential energy is $V(r)$, except that m^* replaces the mass m of the particle.

2·1. *The hydrogen atom*

Equation (6) shows that the quantised values of the internal energy W_2 of an atom consisting of a nucleus of mass M and charge Ze and an electron are

$$W = -\frac{m^* Z^2 e^2}{2n^2 \hbar^2},$$

where $m^* = m/(1+m/M)$. e, h, and m are scarcely known accurately enough for the difference between m and m^* to be observable directly in the spectrum of hydrogen; but the difference in the Rydberg constants as deduced from the spectra of hydrogen ($Z = 1$) and ionised helium ($Z = 2$) enables the effect to be seen and a value of m/M obtained in good agreement with that from other sources.†

† See, for instance, J. A. Crowther, *Ions, Electrons and Ionising Radiations*, 7th ed., London, 1944. p. 274.

2·2.   *The diatomic molecule*

We have here at least three particles to consider: the two nuclei, and one or more electrons. In what follows we shall as usual denote the mass of each electron by m and the co-ordinates of all of them by the single symbol q. The masses of the two nuclei will be denoted by M_1, M_2 and their coordinates by

$$\mathbf{R}_1 = (X_1, Y_1, Z_1)$$

and $$\mathbf{R}_2 = (X_2, Y_2, Z_2).$$

We write the distance between the nuclei as

$$R = |\mathbf{R}_1 - \mathbf{R}_2|.$$

It was first shown by Born and Oppenheimer† that it is permissible, to a good approximation, to treat the molecule in the following way. First we solve (or imagine solved) the Schrödinger equation for the electrons moving in the field of the nuclei supposed at rest at the points $\mathbf{R}_1$, $\mathbf{R}_2$; the solutions will be a series of wave functions $\psi_n(\mathbf{R}_1, \mathbf{R}_2; q)$ with corresponding energy values $W_n(R)$. We then treat this energy $W_n(R)$ as though it were part of the potential energy of the two nuclei when distant R apart. In fact we take for this potential energy

$$V(R) = \frac{Z_1 Z_2 e^2}{R} + W_n(R), \qquad (7)$$

thus adding to $W_n(R)$ the Coulomb interaction of two nuclei with charges $Z_1 e$, $Z_2 e$.

A discussion of some methods of calculating $V(R)$ will be given in § 6; we obtain, for normal and excited states, curves such as shown in Fig. 21; for a stable molecule the ground state has a minimum, shown at the point P. The value R_0 for which this occurs is the equilibrium value of the distance between the nuclei.

† M. Born and J. R. Oppenheimer, *Ann. Phys., Lpz.* LXXXIV, 457, 1927.

Following the method of Born and Oppenheimer, we then write down the Schrödinger equation for a pair of nuclei acting on each other with a force such that their potential energy is $V(R)$. This is

$$\frac{\hbar^2}{2M_1}\nabla_1^2\psi + \frac{\hbar^2}{2M_2}\nabla_2^2\psi + \{W - V(R)\}\psi = 0. \tag{8}$$

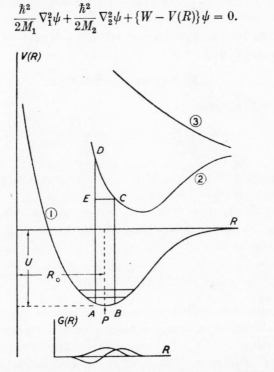

Fig. 21. Energies $V(R)$ of a diatomic molecule as a function of the distance R between the nuclei; curve (1) is for the ground state, curves (2) and (3) for excited states. The vibrational wave functions for the first two excited states are shown below.

Just as for the hydrogen atom the wave function can be separated into a factor f describing the motion of the centre of gravity, and a factor $g(\mathbf{R})$ describing the behaviour of the line joining the nuclei, that is to say, the vector $\mathbf{R}$. It is this second function $g(\mathbf{R})$ which is of interest; it describes *rotation*

of the molecule about its centre of gravity, and *vibration* about
the mean position P. The equation satisfied by g is, writing M
for $M_1 M_2/(M_1 + M_2)$,

$$\frac{\hbar^2}{2M} \nabla^2 g + \{W - V(R)\} g = 0, \qquad (9)$$

which, as for the hydrogen atom, has solutions of the form
(cf. Chap. IV, § 5)

$$g(\mathbf{R}) = P_l^u (\cos \theta) \, e^{iu\phi} G(R)/R,$$

where $G(R)$ satisfies

$$\frac{d^2 G}{dR^2} + \left\{ \frac{2M}{\hbar^2} (W - V) - \frac{l(l+1)}{R^2} \right\} G = 0. \qquad (10)$$

This solution describes a state of the molecule in which it is
rotating with l quanta of angular momentum.

Now the function $V(R)$ has, we assume, a minimum for
$R = R_0$. In the neighbourhood of this value of R, then, we
may write

$$V(R) = U \left\{ -1 + \frac{\alpha(R - R_0)^2}{R_0^2} \right\}, \qquad (11)$$

where α is a dimensionless constant, in general of order unity.
U is the dissociation energy of the molecule, for most molecules
in the range 2–6 eV. U is thus of the order of the excitation
potential of an atom, and R_0 of the atomic radius. But M is
several thousand times greater than the electronic mass. Thus
the arguments of Chapter IV, § 3, show that the radial extension
of the wave function G, for the first few states at any rate, is
small compared with R_0. It is thus legitimate to use the
approximation (11) for $V(R)$, and also to replace R by R_0 in
the term $l(l+1)/R^2$ in (10). Equation (10) thus becomes

$$\frac{d^2 G}{dR^2} + \left\{ \frac{2M}{\hbar^2} (W + U) - \frac{l(l+1)}{R_0^2} - \frac{2MU\alpha}{\hbar^2 R_0^2} (R - R_0)^2 \right\} G = 0.$$

Comparing this with (4) of Chapter IV, we see that x may be replaced by $R - R_0$ and p by $2U\alpha/R_0^2$; the quantised values of the energy W are given by

$$W = -U + \frac{\hbar^2 l(l+1)}{2MR_0^2} + h\nu(n + \tfrac{1}{2}), \qquad (12)$$

where n is an integer and

$$\nu = \frac{1}{2\pi}\sqrt{\frac{p}{M}} = \frac{1}{2\pi}\sqrt{\frac{2U\alpha}{MR_0^2}}.$$

Various points may be noted about this formula. The term $\hbar^2 l(l+1)/2MR_0^2$ represents the rotational energy of a molecule with moment of inertia MR_0^2 and l quanta of angular momentum; it replaces the formula $\hbar^2 l^2/2MR_0^2$ [Chap. IV (1)] of the old quantum theory. It will be noticed that the interval between energy levels is smaller by a factor of order m/M than the interval between energy levels of an atom. The last term in (12) gives the vibrational energy. The interval $h\nu$ between vibrational levels is of order $\sqrt{(M/m)}$ larger than that between rotational levels. The energy levels of a molecule are thus crowded far more closely together than those of an atom and give rise to the so-called band spectra.

2·3.   *The Franck-Condon principle*

This states that if a molecule is in a vibrational state such as that represented by the horizontal line AB in Fig. 21, and if through absorption of radiation it makes a transition to an excited electronic state such as that marked (2) in the same figure, then the energy of the quantum absorbed will normally lie between AD and BC. It is supposed that the atomic nuclei are vibrating slowly between A and B, and that the change in the electronic configuration is rapid compared with the nuclear motion.

It will be seen that the energy $h\nu$ required to make an electron jump to an excited electronic state in a molecule is in general greater than the minimum energy required to reach the excited state.

Exercises

(1) Prove that at temperatures such that $kT \gg h\nu$ the width (*ED* in Fig. 21) of the absorption band of a molecule corresponding to a given electronic transition is proportional to $T^{\frac{1}{2}}$.

(2) Obtain a solution of the Schrödinger equation for the potential energy function introduced by Morse†

$$V(R) = \beta\{1 - e^{-\alpha(R-R_0)}\}^2$$

for the case $l = 0$.

(3) How much does the value of l affect the equilibrium value R_0 of R, and the vibrational frequency ν? (In the approximation of § 2·2 there is no change with l.)

(4) If the potential energy function near the minimum is

$$V(R) = -U + \alpha(R - R_0)^2 + \beta(R - R_0)^3,$$

use perturbation theory (Chap. IV, § 9) to evaluate the effect of the final term on the energy of the first and second vibrational states. In practice how great do you expect this correction to be?

3. SYSTEMS CONTAINING TWO OR MORE PARTICLES OF THE SAME TYPE

In this section we shall discuss certain important properties of systems containing two or more particles of the same type. Examples of such systems are the helium atom (two electrons), the hydrogen molecule (two protons, as well as two electrons), the oxygen molecule (two oxygen nuclei).

We shall begin by confining ourselves to systems containing two such particles. We shall first prove, as a theorem in mathematics, that if the system is in a non-degenerate stationary state with quantised energy, the wave function is either symmetrical or antisymmetrical in the coordinates of the two particles.

By this we mean the following. Let q_1 denote the spatial coordinates (x_1, y_1, z_1) of one of the particles together with its

† P. M. Morse, *Phys. Rev.* xxxiv, 57, 1929.

spin coordinate σ_1, if it has a spin.† Similarly, let q_2 denote the coordinates of the other particle. Then the wave function describing the two particles may be written $\psi(q_1, q_2)$. The wave function is said to be‡ symmetrical if

$$\psi(q_1, q_2) = \psi(q_2, q_1),$$

and antisymmetrical if

$$\psi(q_1, q_2) = -\psi(q_2, q_1).$$

The proof is as follows. For brevity we write 1 for q_1, 2 for q_2. If the wave equation is written

$$\{H(1, 2) - W\}\psi(1, 2) = 0, \tag{13}$$

then the operator $H(1, 2)$ is *necessarily* symmetrical; i.e.

$$H(1, 2) = H(2, 1). \tag{14}$$

This follows from the fact that the particles are identical. Suppose, then, we interchange 1 and 2 in equation (13); we obtain

$$\{H(2, 1) - W\}\psi(2, 1) = 0,$$

and by (14) this may be written

$$\{H(1, 2) - W\}\psi(2, 1) = 0.$$

It follows that $\psi(2, 1)$ is a solution of the original wave equation (13). But we have already stated that $\psi(1, 2)$ is a non-degenerate solution; that is to say, for the energy W there is no other solution satisfying the boundary conditions. Thus $\psi(2, 1)$ must be a multiple of $\psi(1, 2)$, so that

$$\psi(2, 1) = A\psi(1, 2). \tag{15}$$

Interchanging 1 and 2, this gives

$$\psi(1, 2) = A\psi(2, 1). \tag{16}$$

---

† Electrons, protons, and neutrons have a spin, alpha-particles do not.
‡ Thus the function

$$\mathbf{r} = \sqrt{\{(x_1 - x_2)^2 + (y_1 - y_2)^2 + (z_1 - z_2)^2\}}$$

is symmetrical, $\cos\theta = (z_1 - z_2)/r$ antisymmetrical, and functions such as $(1 + \cos\theta)/r$ unsymmetrical.

Multiplying these equations together and dividing by

$$\psi(1, 2)\,\psi(2, 1),$$

we find $$A^2 = 1,$$

whence $$A = \pm 1.$$

It follows that all non-degenerate quantised solutions are either symmetrical or antisymmetrical.

The simplest example is the idealised problem of a diatomic molecule rotating in two dimensions, i.e. in a plane; the rotation is described by the wave functions $e^{\pm il\theta}$, where θ is the angle which the line joining the nuclei makes with a fixed axis. Since θ changes to $\pi + \theta$ when the positions of the particles are interchanged, it will easily be seen that wave functions for even values of l are symmetrical, those for odd values antisymmetrical. The same is true for the wave functions for rotation in three dimensions, $P_l^u(\cos\theta)\,e^{iu\phi}$.

We shall next prove that, if a system containing two identical particles is in a state described by a symmetrical wave function, it can *never* make a transition to a state described by an antisymmetrical wave function, and vice versa. This follows at once from the wave equation which determines the rate of change of ψ, namely (cf. p. 46)

$$-\frac{\hbar}{i}\frac{\partial\psi}{\partial t} = H(1, 2)\,\psi(1, 2).$$

Since $H(1, 2)$ is symmetrical, the change $\delta\psi$ in ψ which takes place in time δt must have the same symmetry as ψ. Thus a symmetrical function will stay symmetrical and an antisymmetrical one will stay antisymmetrical, for all time and under any perturbation whatever.

We must now appeal to experiment, and state that for electrons, protons, and neutrons quantised states with antisymmetrical wave functions are the only states found in nature, while for certain nuclei, notably He^4, C^{12} and O^{16}, only those with symmetrical wave functions† are found. The evidence for this will be given below; we discuss first, however, the reason why only half the theoretically possible states occur in nature.

Suppose that two particles are under consideration, and that measurements are made of the position, momentum, and spin direction of one, which can be represented in the sense of Chapter III, § 3, by a wave function $w_a(q)$, and also of the other, which can be represented by a wave function $w_b(q)$. Then it might seem natural to set for the wave function $\psi(q_1, q_2)$ of the pair of particles

$$w_a(q_1)\, w_b(q_2).$$

Then $\qquad\qquad | w_a(q_1)\, w_b(q_2) |^2 dq_1 dq_2$

would be equal to the chance of the first particle having co-ordinates between q_1 and $q_1 + dq_1$ and the second between q_2 and $q_2 + dq_2$. But such a wave function tells us more about the system than we can in fact know about it. If the particles are of the same type, it is impossible to tie a label on to one of them, and to say that this is the 'first particle', and it is this one that we find at q_1. The wave function $\psi(q_1, q_2)$ should give, when one writes down and interprets its square $| \psi |^2$, the chance that at q_1 one particle will be found, and at q_2 the other, without making any statement about which is which. If this is the correct interpretation of the wave function, it is clear that

$$| \psi(q_1, q_2) |^2$$

must be symmetrical. Further, this will be the case if we set for ψ

$$\psi = w_a(q_1)\, w_b(q_2) \pm w_a(q_2)\, w_b(q_1),$$

but not for the simple products. Unless, however, we make some hypothesis as to whether the wave functions should be symmetrical or antisymmetrical, an ambiguity exists in wave mechanics; we do not know whether to take the plus or minus sign.

† Particles for which the wave functions are symmetrical are said to obey Einstein-Bose statistics, those for which they are antisymmetrical Fermi-Dirac statistics.

4. APPLICATIONS OF THE SYMMETRY PROPERTIES OF THE WAVE FUNCTIONS

4·1. *The exclusion principle*

This important principle, first introduced into wave mechanics by Pauli, states that no two electrons in a given atom can have the same quantum numbers. The validity of this principle shows at once that wave functions describing electrons are antisymmetrical. For suppose two electrons, to the approximation in which we can neglect their interaction, have the wave functions $w_a(q)$, $w_b(q)$. The antisymmetrical function formed from these is

$$w_a(q_1)\, w_b(q_2) - w_a(q_2)\, w_b(q_1).$$

But if $w_a(q)$, $w_b(q)$ are the same function, as they will be if both electrons are in the same state, this must vanish. Therefore it is not possible for both particles to be in the same state, if the function is antisymmetrical.

The wave function w includes the spin coordinate; a more detailed discussion of the spin coordinates is given in § 4·3.

4·2. *Particles without spin obeying Einstein-Bose statistics*

If a particle has no spin, it is described by the spatial or orbital coordinate $\mathbf{r} = (x, y, z)$ only. Certain nuclei such as He^4, C^{12} and O^{16}, already mentioned as obeying Einstein-Bose statistics, have no spin. We consider rotational states of molecules containing two of these nuclei (e.g. O_2). The rotational wave functions of a diatomic molecule are symmetrical† in the coordinates $\mathbf{r}_1, \mathbf{r}_2$ of the two nuclei for even values of the quantum number l describing the rotation of the molecule, antisymmetrical for odd values of l. Analysis of the band spectra shows that only states with even values of l are found for O_2 and for the (unstable) molecule He_2. This shows that the nuclei have no spin and obey Einstein-Bose statistics (i.e. have symmetrical wave functions).

---

† This is true for the lowest electronic state; for certain excited states it may be the other way round, as may be seen from the arguments of § 6·3.

Other evidence can be derived from the scattering of α-particles in helium, where the symmetry of the wave function describing the two particles leads to anomalies in the scattering.†

4·3. *Particles with spin obeying Fermi-Dirac statistics*

It has been emphasised in Chapter IV, § 13, that the effect of the spin on the orbital wave functions is small, and that to a good approximation one can represent the wave function of a particle with spin by a product of the form

$$\psi(x, y, z)\chi(\sigma).$$

σ is here ± 1, according as the spin direction lies parallel or antiparallel to a fixed axis, and χ is capable of taking two independent forms χ_α and χ_β. In the same way, then, the wave function of a pair of particles with spin can be written

$$\psi(\mathbf{r}_1, \mathbf{r}_2)\chi(\sigma_1, \sigma_2),$$

where ψ is a solution of an equation of the type (3) in which the spin is neglected. The symbol $\mathbf{r}_1$ here denotes x_1, y_1, z_1, etc. Now solutions of this equation, for non-degenerate quantised states, are either symmetrical or antisymmetrical in $\mathbf{r}_1, \mathbf{r}_2$; we may write them $\psi_S(\mathbf{r}_1, \mathbf{r}_2), \psi_A(\mathbf{r}_1, \mathbf{r}_2)$. Thus the wave function of the whole system, which must be antisymmetrical for an interchange of q_1 (denoting the whole group x_1, y_1, z_1, σ_1) and q_2, must have one or other of the forms

$$\psi_S(\mathbf{r}_1, \mathbf{r}_2)\chi_A(\sigma_1, \sigma_2), \quad \psi_A(\mathbf{r}_1, \mathbf{r}_2)\chi_S(\sigma_1, \sigma_2). \tag{17}$$

where χ_S, χ_A are themselves symmetrical and antisymmetrical in σ_1, σ_2. Now such functions can be formed from χ_α, χ_β only as follows:

A symmetrical function $\chi_S(\sigma_1, \sigma_2)$ can be formed in *three* ways:

$$\left.\begin{array}{l} \chi_\alpha(1)\chi_\alpha(2), \\ \chi_\beta(1)\chi_\beta(2), \\ \chi_\alpha(1)\chi_\beta(2) + \chi_\alpha(2)\chi_\beta(1), \end{array}\right\} \tag{18}$$

† Cf. Mott and Massey, p. 102.

and an antisymmetrical function $\chi_A(\sigma_1, \sigma_2)$ in only one way:

$$\chi_\alpha(1)\chi_\beta(2) - \chi_\alpha(2)\chi_\beta(1).$$

We thus reach the following conclusion. In any system containing two particles with spin, the states can be separated into those with symmetrical orbital wave functions ψ_S, including usually the ground state, and those with antisymmetrical orbital functions ψ_A. Transitions between states with symmetrical and antisymmetrical orbital wave functions, though possible, will have very low probability; to the approximation that the wave function (17) is valid, they do not occur. All states with symmetrical orbital wave functions are singlet states, the two spins being antiparallel so that they contribute nothing to the mechanical or magnetic moment; all states with antisymmetrical orbital functions are triplets, the total spin moment along a fixed direction being, in multiples of $\hbar$, -1, 0, or $+1$.

4·4. The rotational states of H_2

The proton has a spin with angular momentum $\tfrac{1}{2}\hbar$ and a magnetic moment of about two nuclear magnetons $(e\hbar/2Mc)$. Owing to the smallness of this magnetic moment, which will influence the orbital wave functions very little, the transition probabilities from rotational states for which l is even (symmetrical wave functions) to those for which l is odd will be very small indeed. In fact for ordinary purposes hydrogen gas can be regarded as a mixture of two gases, parahydrogen (molecules in singlet states for which l is even) and orthohydrogen (molecules in triplet states for which l is odd). Transitions from one state to the other take place practically only in the presence of a catalyst which dissociates the molecules into atoms, so that an atom from one molecule can recombine with that from another.[†] In the absence of a catalyst, when the gas is cooled the numbers of molecules in the various rotational states will not reach the equilibrium values, because molecules cannot

[†] Cf. A. Farkas and L. Farkas, *Proc. Roy. Soc.* A, cliii. 124, 1935; D. D. Eley, 'The Catalytic Activation of Hydrogen', contribution to *Advances in Catalysis*, New York, 1949, vol. 1, p. 157.

jump from the state $l = 1$ to $l = 0$. This fact has an important effect on the specific heat at low temperatures, which has been observed experimentally.[†]

4·5. *The helium atom*

The Schrödinger equation for two electrons moving in the field of a nucleus is

$$(H - W)\psi = 0,$$

where

$$H\psi = -\frac{\hbar^2}{2m}(\nabla_1^2 + \nabla_2^2)\psi + V\psi, \qquad (19)$$

and where

$$V = -\frac{2e^2}{r_1} - \frac{2e^2}{r_2} + \frac{e^2}{|\mathbf{r}_1 - \mathbf{r}_2|}.$$

The first two terms in V represent the potential energy of the two electrons in the field of the nucleus, and the final term the interaction energy of the two electrons.

It follows from the arguments of § 4·3 that the solutions of this equation are either symmetrical or antisymmetrical: that states with symmetrical orbital functions are singlets (known as parahelium states), that states with antisymmetrical orbital functions are triplets (known as orthohelium), and that transitions from one series to another have very low probability. In this section we shall show how to calculate approximately the energies of these states.

We shall start with the approximation in which each electron is given a separate orbital wave function, denoted by $\psi_a(r)$, $\psi_b(r)$. We shall suppose, moreover, that these functions are either the same ($\psi_a = \psi_b$), or else that they are orthogonal. Then an approximate wave function describing the pair of electrons will be

$$\psi(1, 2) = \frac{1}{\sqrt{2}}\{\psi_a(1)\psi_b(2) \pm \psi_a(2)\psi_b(1)\}. \qquad (20)$$

[†] Cf. the review by R. H. Fowler. *Statistical Mechanics*, 2nd ed., Cambridge, 1936, p. 82.

The factor $1/\sqrt{2}$, if ψ_a and ψ_b are normalised, ensures that the wave function $\psi(1, 2)$ is normalised, in other words that it satisfies the equation

$$\iint |\psi(1, 2)|^2 d\tau_1 d\tau_2 = 1.$$

Approximate values of the energy of the atom can then be found from the formula

$$W = \iint \psi^* H \psi d\tau_1 d\tau_2,$$

W being here the total energy of both electrons.

Substituting for $\psi(1, 2)$ we find

$$W = I \pm J, \tag{21}$$

where
$$\left.\begin{aligned} I &= \iint \psi_a^*(1) \psi_b^*(2) H \psi_a(1) \psi_b(2) \, d\tau_1 d\tau_2, \\ J &= \iint \psi_a^*(1) \psi_b^*(2) H \psi_a(2) \psi_b(1) \, d\tau_1 d\tau_2. \end{aligned}\right\} \tag{22}$$

The integral J is known as an 'exchange' integral.

On the basis of this theory we can give a descriptive analysis of the level scheme of helium: for the ground state we may set $\psi_a = \psi_b$, both being spherically symmetrical functions of the type $Ce^{-r/a}$. Then only a symmetrical function of the type $\psi_a(r_1)\psi_a(r_2)$ is possible. But if one of the electrons is excited, there exist *two* states with different energy levels, corresponding to the two signs in (20) and (21). Moreover the ground state and the excited states which have symmetrical wave functions in the spatial coordinates $\mathbf{r}_1, \mathbf{r}_2$ are singlet states (known as states of parahelium); the spin wave function is of the form

$$\chi_\alpha(\sigma_1) \chi_\beta(\sigma_2) - \chi_\alpha(\sigma_2) \chi_\beta(\sigma_1),$$

and so the two spins point in opposite directions and make no contribution to the mechanical and magnetic moments. The ground state, for which both electrons have spherically symmetrical wave functions, has thus no resultant angular momentum or magnetic moment at all; excited states, such as that in which one electron remains in the $1s$ state while the other is in a state with orbital angular momentum $l\hbar$, have angular momentum of this amount and magnetic moment $\mu_B l$. They show a normal Zeeman effect.

The states with antisymmetrical orbital wave functions (states of orthohelium) have the three spin wave functions

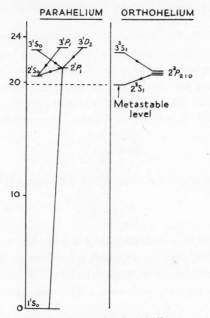

Fig. 22. Energy levels of helium.

given by (18). They are thus triplet states, of total spin moment *one* unit of $\hbar$. Provided that the excited electron is not in an s state ($l = 0$), the spin magnetic moment will interact with the orbital moment so that the state will split into three states with different energies. States in which the excited electron is in an s state will not be split except in the presence of a magnetic field.

A schematic representation of the energy levels of helium is shown in Fig. 22.

We may note here† that the exchange integral is positive. Thus the triplet levels (orthohelium) lie below the singlet levels. This may be understood, because an antisymmetrical orbital wave function must vanish when $\mathbf{r}_1 = \mathbf{r}_2$ and will in general be small when the particles are close together. Thus the *positive* contribution to the energy made by the interaction term $e^2/|\mathbf{r}_1 - \mathbf{r}_2|$ is smaller than for the symmetrical states.

4·6. *The structure of atoms with more than two electrons*

In describing atoms more complicated than helium, the most convenient approximation is the following:

All electrons are supposed to move in the so-called self-consistent field. This field, in which the potential energy of an electron is $V(r)$, is defined as follows. For small r

$$V(r) \sim -\frac{Ze^2}{r},$$

where Z is the atomic number, and for large values of r

$$V(r) \sim -\frac{e^2}{r}.$$

At other points it is defined as the field of the nucleus and the *average* field produced by all the other electrons, if each of them is treated as producing a charge density $-e\,|\psi(r)|^2$. It will be noticed that with this definition $V(r)$ is not quite the same for each electron.

In this field the lowest level, with principal quantum number $n = 1$, is known as the K level. The energy of such a level is given approximately by the Bohr formula

$$-me^4Z^2/2\hbar^2;$$

† For the evaluation of the exchange integrals and suitable choice of the wave functions ψ_a, ψ_b, the reader is referred to the following authorities:

(i) The original paper on the subject is by W. Heisenberg, Z. *Phys.* xxxix, 499, 1926.

(ii) A very complete account of the calculations made up to 1933 is by H. Bethe, *Handb. Phys.* xxiv, pt. 1, 342, 1933.

(iii) L. Pauling and E. B. Wilson, *Introduction to Quantum Mechanics*, New York, 1935, p. 210.

for $Z = 79$ (gold), for instance, this is 84,000 eV. By the Pauli principle two electrons can be accommodated in the K level.

The next set of levels, with principal quantum number 2, can accommodate eight electrons, two in s states $(l = 0)$ and six in p states $(l = 1)$. The electrons in these states form what is known as the L shell. There are in fact three L levels; the level

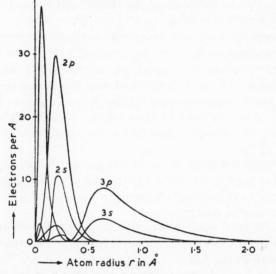

Fig. 23. Radial charge distribution for the different electron groups of K^+.

L_I, with quantum number $l = 0$ $(2s)$, which can accommodate two electrons, and the levels L_{II} and L_{III}, with $l = 1$ $(2p)$, and spin directions either parallel or antiparallel to the direction of the orbital momentum.

Owing to the Pauli principle, an electron cannot normally jump from an L to a K level, since all the states in the K level are occupied. Under electron bombardment, however, a K electron may be ejected from an atom; an L electron may then fall into the vacant level, a quantum of X-radiation being emitted. Processes of this type are responsible for the emission of X-ray lines. The transitions from the L_{II} and L_{III} levels to the K level give rise to the lines known as K_{α_1} and K_{α_2}.

Fig. 23 shows the charge distribution calculated by the method of the self-consistent field in the several shells of the potassium positive ion. The quantity plotted here is $r^2 |\psi|^2$, and thus the number of electrons between spheres of radii $r, r+dr$, divided by $4\pi dr$.

Some points to notice about the shell distribution of electrons in atoms are the following:

(1) Normally only the outermost shell of electrons in an atom is responsible for chemical binding, and is thus affected by the chemical state of an atom. Therefore the frequencies and breadths of most X-ray emission lines, unlike optical spectra, are almost unaffected by the state of chemical binding. Breadths and structures of lines which start from the outermost level do, however, depend on the chemical state of the atom (cf. § 7).

(2) Any closed shell, K, L or M, i.e. one in which all the states of a given principal quantum number are occupied, has a spherically symmetrical charge distribution, and no resultant spin or magnetic moment.

The reader will easily verify that for an L shell, for instance, the distribution is spherically symmetrical; since the three p functions have the form

$$xf(r), \quad yf(r), \quad zf(r),$$

the resultant charge density is

$$-2e(x^2+y^2+z^2)\{f(r)\}^2.$$

(3) Information about the charge density in atoms can be obtained experimentally from the intensities with which atoms (for example in a crystal) scatter X-rays. The argument is as follows. If a polarised light wave falls on a free electron, the classical formula for the scattered amplitude, measured at a point P at distance R from it, is†

$$\frac{E}{R}\frac{e^2}{mc^2}\sin\phi;$$

† Cf. Chap. IV, § 10.

here ϕ is the angle between the electric vector E of the incident wave and the line joining the electron to the point P. This formula is not in fact valid for *free* electrons, for which the Compton effect with recoil of the electron is to be expected: but the formula may be used to calculate the coherent scattering by an atom of radiation for which the frequency is great compared with the absorption frequencies of the atom. One assumes that if $e\,P(r)$ is the charge density, then each element of volume $d\tau$ scatters a wavelet of amplitude

$$\frac{E}{R}\frac{e^2}{mc^2}P(r)\,d\tau\sin\phi,$$

and that these wavelets interfere. The resultant amplitude may be calculated exactly as in Chapter II, § 7; corresponding to formula (12) the resultant amplitude is

$$\frac{E}{R}F(\theta)\sin\phi, \tag{23}$$

where
$$F(\theta) = \int_0^\infty P(r)\frac{\sin\left(4\pi r\sin\theta/\lambda\right)}{4\pi r\sin\theta/\lambda}\,4\pi r^2 dr.$$

2θ is here the angle of scattering and λ the wavelength of the X-radiation.† $F(\theta)$ is known as the atomic scattering factor.

The intensity of an unpolarised wave is proportional to $\frac{1}{2}(1+\cos^2\theta)$ instead of $\sin^2\phi$.

(4) The diamagnetic susceptibility depends critically on the radial extent of the wave function; it is given by‡

$$\chi = \frac{Ne^2}{6mc^2}\int\sum_s r_s^2\,|\,\psi(\mathbf{r}_1,\mathbf{r}_2,...,\mathbf{r}_n)\,|^2 d\tau_1...d\tau_n,$$

† For a review of this subject, see, for example, R. W. James, *The Optical Principles of the Diffraction of X-rays*, London, 1948, chap. III. For some recent determinations of the electron density in the ions of alkali-halide crystals, see R. Brill, H. G. Grimm, C. Hermann and C. Peters, *Ann. Phys., Lpz.*, XXXIV, 393, 1939, or the review of this work by A. Eucken, *Lehrbuch der chemischen Physik*, 1944, vol. II, pt. 2, p. 537.

‡ Cf. E. C. Stoner, *Magnetism and Matter*, Leipzig, p. 108.

where N is the number of atoms per unit volume, and the summation is over all the n electrons of the atom.

A number of approximate methods exist for obtaining wave functions of atoms. There is first of all the method of the self-consistent field due to Hartree, already mentioned. Then various improvements to the Hartree method, notably that of Fock, have been developed.[†] Finally the Thomas-Fermi[‡] method is available for atoms too complicated to be treated by other methods.

5. INTERATOMIC FORCES AND THE FORMATION OF MOLECULES

Forces between atoms are of the following types:

(a) *Van der Waals attractive forces.*[§] A calculation based on wave mechanics shows that, at sufficiently large distances r, all atoms and molecules attract each other with a force of which the potential energy $V(r)$ is of the form $-C/r^6$. This force is of importance in considering the equation of state of imperfect gases, and is responsible for cohesion in solid or liquid rare gases, methane (CH_4), solid hydrogen, etc. The constant C is given in terms of the absorption frequencies ν_r and the related oscillator strengths f_r.

If we make the approximation that the oscillator strengths of all lines can be neglected except those of one line for each atom, of frequencies ν, ν', then

$$C = \tfrac{3}{2}h\left(\frac{e^2}{4\pi^2 m}\right)^2 \frac{1}{\nu\nu'(\nu+\nu')}.$$

For the rare gases, where no other attractive force comes into play, the van der Waals attraction is rather weak, on account of the high values of the excitation potentials. This is shown by the low values of the melting and boiling points of these substances.

---

† A review of these methods and of the results obtained have been given by D. R. Hartree, *Rep. Prog. Phys.* II, 113, 1946-7.

‡ Cf., for example, Mott and Sneddon, p. 158.

§ For further details compare, for example, Mott and Sneddon, p. 164.

The van der Waals force is the only force between neutral atoms when their charge clouds do not overlap.

(b) *Repulsive overlap forces.* As soon as two closed shells overlap, a strong repulsive force sets in; in principle this can be calculated by the methods of wave mechanics, but the calculations are laborious and the results obtained not always reliable;† it is usual to use empirical forms such as

$$Ar^{-s} \quad (s \sim 9-12), \quad \text{or} \quad Be^{-\mu r}$$

for the potential energy of two atoms distant r apart.

(c) *Ionic forces.* It seems to be a good approximation to treat alkali-halide crystals as made up of positive metal ions and negative halide ions, the crystal being held together by the electrostatic attraction between them, and the ions kept apart by the repulsive overlap forces. Much work has been done in explaining in terms of these forces the properties of crystals built of atoms or ions of which the outermost electronic shell is closed; e.g. alkali-halides and solid rare gases. This, however, is not strictly a part of wave mechanics, and we shall confine ourselves to giving references here.‡

6. COVALENT FORCES

Apart from the van der Waals forces and electrostatic attraction between ions, forces between atoms arise only when the wave function of one atom overlaps that of the next. The valence forces of chemistry are of this type.

† For a recent attempt, cf. G. Wyllie and E. F. Benson, *Proc. phys. Soc.* A, LXIV, 276, 1951.

‡ For alkali-halides and crystals held together by ionic forces, see the following: M. Born and M. Göppert-Mayer, *Handb. Phys.* XXIV, pt. 2, 625, 1933; N. F. Mott and R. W. Gurney, *Electronic Processes in Ionic Crystals*, 1940, chap. I; J. Sherman, *Chem. Rev.* XI, 153, 1932.

For criticism of the approach under (c) above, and the refinements introduced by wave mechanics, see P. O. Löwdin, *A Theoretical Investigation into some Properties of Ionic Crystals*, Upsala, 1949.

For a discussion of solid rare gases, see J. E. Lennard-Jones and B. M. Dent, *Proc. Roy. Soc.* A, CXIII, 673, 1927.

We shall discuss valence forces in the following way:

(*a*) We shall give a treatment of the movement of an electron in the hydrogen molecular ion H_2^+, where the problem is that of the movement of a single electron in the field of two protons.

(*b*) We shall discuss the hydrogen molecule H_2 by a method known as the method of molecular orbitals,† in which each electron is pictured as shared between the two atoms in the same way as in H_2^+, so that each electron can move independently from one atom to the other.

(*c*) An outline will be given of an alternative mathematical approach, that of London-Heitler, in which each electron is located on its own atom; they are allowed to change places, but not to move across independently of each other.

It will be realised that (*b*) and (*c*) are alternative approximations; the true wave function will lie somewhere between the two extremes.

(*d*) Finally we shall enumerate some of the main points in the application of wave mechanics to more complicated molecules.

6·1. *The molecular ion* H_2^+

The essential point in the treatment of this ion is that for every state of the electron in a single atom there will be two states for the molecule. Consider, for example, the $1s$ state of the atom, with a normalised function of the form

$$\psi(r) = Ce^{-r/a},$$

where r is the distance from the nucleus. If the two nuclei are at points defined by the vectors

$$\mathbf{r}_a = (x_a, y_a, z_a),$$

$$\mathbf{r}_b = (x_b, y_b, z_b),$$

† The name is due to J. E. Lennard-Jones, *Trans. Faraday Soc.* xxv. 668, 1929.

then
$$|\mathbf{r} - \mathbf{r}_a| = \sqrt{\{(x - x_a)^2 + (y - y_a)^2 + (z - z_a)^2\}}$$

represents the distance of the electron from the point $\mathbf{r}_a$; a similar expression may be written down for the distance from $\mathbf{r}_b$. Then we may write

$$\psi_a(\mathbf{r}) = \psi(|\mathbf{r} - \mathbf{r}_a|)$$

for an electron in the ground state in atom a, and

$$\psi_b(\mathbf{r}) = \psi(|\mathbf{r} - \mathbf{r}_b|)$$

for an electron in the ground state in atom b.

An electron in the molecule may be located on either of the atoms; one may write the wave function

$$A\psi_a(\mathbf{r}) + B\psi_b(\mathbf{r}).$$

By symmetry it is just as likely to be on one atom as on the other; that is A^2 must be equal to B^2, so that $A = \pm B$ and the two possible wave functions (unnormalised) are

$$\Psi_S = \psi_a(\mathbf{r}) + \psi_b(\mathbf{r}) \tag{24.1}$$

and
$$\Psi_A = \psi_a(\mathbf{r}) - \psi_b(\mathbf{r}). \tag{24.2}$$

Moreover, these have different energies; using the formula for the energy W from Chapter IV, (13), in the form

$$W \int |\Psi|^2 d\tau = \int \Psi^* H \Psi \, d\tau, \tag{25}$$

we see that
$$W(1 \pm \Delta) = I \pm J,$$

where
$$\Delta = \int \psi_a^*(\mathbf{r}) \psi_b(\mathbf{r}) \, d\tau,$$

$$I = \int \psi_a^*(\mathbf{r}) H \psi_a(\mathbf{r}) \, d\tau,$$

$$J = \int \psi_a^*(\mathbf{r}) H \psi_b(\mathbf{r}) \, d\tau.$$

Exercise

Show that $A = \pm B$ by minimising the expression (25) for the energy W.

We shall not discuss the numerical evaluation of the integrals Δ, I, J, since formula (25) gives only an approximate form for the wave function of H_2^+; actually it is possible to obtain an exact solution of the Schrödinger equation for an electron moving in the field of two nuclei.† For our purpose, however,

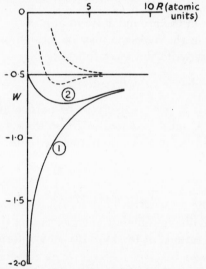

Fig. 24. Energies of an electron in the states of (1) even and (2) odd parity of the ion H_2^+. The full lines represent the energy of the electron, the dotted lines with the addition of the term e^2/R.

we need only notice that the symmetrical solution has no nodal surface; the energy of the electron will tend to $-W_0$ as the distance R between the nuclei tends to infinity, and $-4W_0$ as R tends to zero. W_0 is here the ionisation potential of hydrogen, $me^4/2\hbar^2$. The electron's energy $W(R)$ thus exerts an attractive force between the two nuclei; and it is not surprising

† Cf. O. Burrau, *K. Danske Vidensk. Selsk.* vii, 14, 1927; E. A. Hylleraas, *Z. Phys.* lxxi, 739, 1931; G. Steensholt, *Z. Phys.* c, 547, 1936.

that when the potential energy of the nuclei in one another's field is added to give the total energy $V(R)$,

$$V(R) = W(R) + \frac{e^2}{R}.$$

one obtains a curve with a minimum (Fig. 24). On the other hand the state (24·2) with odd parity tends, as R tends to zero, to the $2p$ state of an electron in the field of a charge $2e$, which has energy $-W_0$. The electron thus exerts no attractive force; and when the term e^2/R is added the potential energy leads to repulsion only.

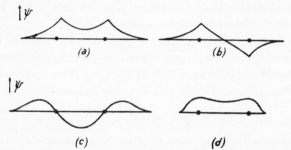

Fig. 25. Wave functions of the hydrogen molecular ion; (a), (b), and (c) are plotted along the line joining the nuclei; (a) and (b) show the two states derived from s atomic wave function, (c) and (d) those derived from p atomic wave function.

The considerations of Chapter VI show that optical transitions are allowed between the two states considered here.

The states formed from atomic wave functions of p symmetry are also of interest. If the nodal surface is perpendicular to the line joining the nuclei, the wave function corresponding to the lower of the two states will be as in Fig. 25c. If, on the other hand, the nodal surfaces are in the plane of the paper, the wave function will change sign in going through the paper, but when they are plotted along any line parallel to the line joining the nuclei the wave function will be as in Fig. 25d. States for which a nodal surface passes through the line joining the nuclei are called π states; those for which the line joining the nuclei contains no node are called σ states.

6·2. *The method of molecular orbitals for the hydrogen molecule* H_2

In this method the two electrons in the molecule are both described by the same wave function $\Psi(\mathbf{r})$, namely, one of the type (24·1) illustrated in Fig. 25a. The wave function of the pair of electrons in the ground state is, including spin coordinates,

$$\Psi(\mathbf{r}_1)\,\Psi(\mathbf{r}_2)\,\{\chi_\alpha(1)\,\chi_\beta(2) - \chi_\alpha(2)\,\chi_\beta(1)\}.$$

The ground state, like the ground state of the helium atom, thus has no magnetic moment due to its electrons.

Excited states can be treated in a similar way. As for helium, there is a singlet and a triplet series of terms.†

The error in the method of molecular orbitals is that it neglects altogether any correlation between the positions of the electrons; it suggests that it is as likely that both electrons are located on the same atom as that they are on different atoms.

6·3. *The method of London-Heitler as applied to* H_2

This method of approach goes to the opposite extreme, and starts with an approximation in which the two electrons are located definitely on different atoms. The electrons are allowed to change places, but not to move independently from atom to atom.

The wave function that we use must be antisymmetrical in the coordinates of the two electrons, including spin. The two possibilities, starting with $1s$ wave functions $\psi_a(\mathbf{r}), \psi_b(\mathbf{r})$ for an electron in either atom, are

$$\{\psi_a(1)\,\psi_b(2) + \psi_a(2)\,\psi_b(1)\}\,\chi_A(1,2), \tag{26}$$

$$\{\psi_a(1)\,\psi_b(2) - \psi_a(2)\,\psi_b(1)\}\,\chi_S(1,2). \tag{27}$$

The former is the wave function of the ground state. Here the antisymmetrical spin wave function, as for the helium atom, gives a singlet state, with zero spin moment.

† It must be emphasised that each electronic state is split into a very large number of states by the quantised rotational and vibrational states discussed in § 2·2 above.

The energy W of the two states is, as in (21) and (22),

$$(I \pm J)/(1 \pm \Delta),$$

where I, J are formally as defined on p. 100 and

$$\Delta = \int\int \psi_a^*(1)\,\psi_b^*(2)\,\psi_a(2)\,\psi_b(1)\,d\tau_1 d\tau_2.$$

The wave functions are now not orthogonal and Δ does not vanish, as it does for the analogous case of the helium atom. The integrals can be worked out.† The exchange integral J is negative, and, together with the energy e^2/R of the Coulomb interaction between the nuclei, leads to attraction between the atoms for state (26), repulsion for (27).

In the absence of overlap between the wave functions, I is just equal to the energy of a pair of free atoms, and J and Δ vanish. Cohesion of the type described here is a result of overlap, and in general maximum cohesion occurs when the overlap is a maximum.

6·4. *Some general features of chemical binding*

The considerations of the previous section show

(a) that a bond involving two electrons, one from each atom, can only be formed if the two spins are antiparallel;

(b) that a strong bond involves considerable overlap between the wave functions.

We shall apply these principles first to the water molecule H_2O. The oxygen atom has two electrons in the state $2s$ and four in the state $2p$. Thus two of the $2p$ wave functions are necessarily paired already, while two are ready to form bonds with hydrogen. But these two must be different wave functions; they must necessarily have their nodal surfaces at right angles to each other. That is why the lines joining the nuclei of the two hydrogen nuclei to the oxygen nucleus make approximately a right angle with each other (actually 105°).

† The original paper is that by W. Heitler and F. London, Z. *Phys.* xLIV, 455, 1927. See also reviews of later work by H. Bethe, *Handb. Phys.* xxIV, pt. 1, 535, 1933; and Pauling and Wilson, *Introduction to Quantum Mechanics*, p. 340.

Another important feature of the theory of chemical binding is the occurrence of hybridised wave functions. An example is the behaviour of carbon in diamond or in such compounds as CH_4. Here each carbon atom is surrounded at the four corners of a tetrahedron by atoms with which it forms a homopolar bond. It would not be correct to say of the four outer electrons in carbon that two are in s states and two in p states; it is correct to ascribe to each of them one of four 'hybridised' wave functions, which extend as far as possible in the four tetrahedral directions, so as to give the maximum overlap. These four wave functions are

$$
\left.
\begin{aligned}
\psi_1 &= \tfrac{1}{2}\{\psi(2s) + \psi(2p_x) + \psi(2p_y) + \psi(2p_z)\}, \\
\psi_2 &= \tfrac{1}{2}\{\psi(2s) + \psi(2p_x) - \psi(2p_y) - \psi(2p_z)\}, \\
\psi_3 &= \tfrac{1}{2}\{\psi(2s) - \psi(2p_x) + \psi(2p_y) - \psi(2p_z)\}, \\
\psi_4 &= \tfrac{1}{2}\{\psi(2s) - \psi(2p_x) - \psi(2p_y) + \psi(2p_z)\},
\end{aligned}
\right\}
\tag{28}
$$

The student is recommended to verify that they have the desired properties.

7. THE THEORY OF SOLIDS

In this section there will be space only to summarise a few of the more important contributions made by wave mechanics to the theory of solids.

7·1. *Concept of a conduction band*

Consider any normally non-conducting crystal as, for example, sodium chloride. Suppose that an extra electron is brought from outside and placed on one of the metal ions Na^+. Then according to wave mechanics such an electron will have the following property. It will not stay localised on one ion, but will be able to jump freely from one ion to the next. Moreover, in a perfect lattice, if an electric field F is applied, the electron will be accelerated, the acceleration being given by

$$
m_{\text{eff.}} \ddot{x} = eF.
$$

m_{eff}, the so-called effective mass, may differ somewhat from the mass of a free electron, but is of the same order of magnitude. The concept of a mean free path does not arise for an idealised lattice in which all the ions are held rigidly in position.

This property of an extra electron brought into a crystal can be deduced from the following properties of the solutions of the Schrödinger equation for an electron moving in a *periodic* field. Let $V(x, y, z)$ be the potential energy of an electron moving in the field that it will encounter within the crystal. Then $V(x, y, z)$ will have the same period a as the crystal lattice. The Schrödinger equation is

$$\nabla^2 \psi + \frac{2m}{\hbar^2} \{ W - V(x, y, z) \} \psi = 0.$$

The solutions may be shown to have the form

$$\psi = e^{i(\mathbf{kr})} u_{\mathbf{k}}(x, y, z), \tag{29}$$

where $u_{\mathbf{k}}(x, y, z)$ is periodic, with the same period as the lattice. This may be compared with the form

$$\psi = e^{i(\mathbf{kr})}$$

for a free electron. Thus the wave function of an electron in the lattice is similar to that of a free electron, and represents a particle moving in a definite direction without being scattered. It is modulated, however, by the field of the lattice.

A mean free path arises if the electrons are scattered, and this occurs only if atoms are displaced from their mean position by heat motion, or if impurities are present. In either case the electron wave is scattered. In the case of an atom displaced from its mean position a distance x, the amplitude of the scattered wave is proportional to x, and the intensity is proportional to x^2. Since the mean value of x^2 varies as the temperature, so does the resistance of a metal.

Corresponding to each wave number $\mathbf{k}$ there is an energy value $W(\mathbf{k})$. For small $\mathbf{k}$, W is of the form

$$W(\mathbf{k}) = \hbar^2 k^2 / 2m_{\text{eff}},$$

though for higher energies it is more complicated, and bands of forbidden energy occur. The first band of allowed energies is known as the 'conduction band'.

7·2.   Semi-conductors

Semi-conductors are substances for which the conductivity increases with increasing temperature. Most but not all are activated by impurities; that is to say, their conductivity depends on the presence of small traces of impurity present in concentrations from one part in 10^6 to about one per cent. It is thought that these impurities are dispersed in atomic form, and that each such impurity centre can release an electron with the expenditure of an amount of energy W that is not too big compared with kT, and is small compared with the excitation energy of a free atom. The electrons released are, as we say, 'in the conduction band'; they are free to move through the lattice.

If there are N impurity centres per unit volume, it can be shown that in thermal equilibrium the number n of free electrons (electrons in the conduction band) is per unit volume

$$n = \sqrt{(NN_0)}\, e^{-\frac{1}{2}W/kT},$$

where
$$N_0 = (2\pi mkT/h^2)^{\frac{3}{2}}.$$

The conductivity σ is obtained by multiplying by ev, where v is the mobility,

$$\sigma = nev.$$

Both N_0 and v vary with temperature, but frequently the exponential is the predominating term, so that the slope of a plot of $\ln\sigma$ against $1/T$ enables an estimate of W to be made.

Actually the slope of such a curve decreases with increasing concentration of impurity,† for a reason which is not yet fully understood.

The current theory of the nature of the impurity centres which explains qualitatively why the values of W are so much

† Cf. *The Report on the Reading Conference on Semi-conductors*, London, 1951.

smaller than the ionisation potentials of free atoms, is as follows. An impurity atom can be accepted by a non-metallic substance in various ways. For instance, ZnO accepts excess zinc by taking a zinc ion (Zn^+) into a so-called interstitial position, which means into one of the small gaps between the zinc and oxygen ions of the crystal lattice. Alkali-halides accept excess metal through the presence of lattice sites from which the anion (the halogen ion) is absent. In either case the centre carries a positive charge, so that the field round it is $e/\kappa r^2$, where κ is the dielectric constant of the medium. The centre has to be neutralised by an electron. If an electron is 'in the conduction band', that is to say, free to move from atom to atom, its potential energy in the field of the centre is $-e^2/\kappa r$. It can thus be held in quantised energy levels, exactly similar to those of an electron in the field of a proton, except that in all formula e^2 must be replaced by e^2/κ and m by $m_{\text{eff.}}$. The binding energy is thus

$$W = \frac{m_{\text{eff.}}\, e^4}{2\hbar^2 \kappa^2} = \frac{13 \cdot 5\,\text{eV.}}{\kappa^2}\, \frac{m_{\text{eff.}}}{m}. \tag{30}$$

For materials such as silicon ($\kappa \sim 17$), activated by suitable impurities, this gives, assuming $m_{\text{eff.}}/m \sim 1$, about $0 \cdot 01\,\text{eV}$, in good agreement with experiment.[†]

If the wave function of the bound electron is of the form $Ce^{-r/a}$, the quantity a may be defined as the 'radius' of the orbital. It should be given by

$$a \sim \kappa \hbar^2/m_{\text{eff.}}\, e^2 = \kappa(m/m_{\text{eff.}})\, 0 \cdot 54. \tag{31}$$

The impurity atoms are thus swollen owing to their presence in the dielectric.

It is not suggested that these formulae, (30) and (31), are exact since the assumption that the potential is $-e^2/\kappa r$ cannot be true right up to the dissolved ion. However, they give correctly the order of magnitude.

[†] Cf. G. L. Pearson and J. Bardeen, *Phys. Rev.* LXXV, 865, 1949.

7·3. *Metals*

These are treated by a method similar to the method of molecular orbitals. We shall illustrate this by considering a 'one-dimensional' metal, i.e. a row of N atoms, each containing one electron. It is convenient to think of this row as bent into a closed loop (Fig. 26), round which the electrons can move.

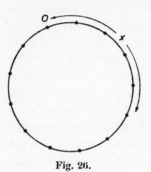

We then ascribe to each electron a wave function of the type

$$e^{ikx}u_k(x), \qquad (32)$$

where x denotes the distance of the point considered from a fixed point O on the loop. In order that the wave function (32) may join up smoothly again at O, k must satisfy the quantum condition

Fig. 26.

$$k = \pm 2\pi n/L,$$

where L is the circumference and n is an integer. We now introduce the Pauli principle, that only two electrons, with spins in opposite directions, can be in states described by the wave function with given k. Thus states are occupied with all integral values of n between $\pm \frac{1}{4}N$; higher states, at the absolute zero of temperature, are empty. The electrons have kinetic energies between zero and $W_{\max.} = \pi^2\hbar^2(N/L)^2/8m$. Since (N/L) is the interatomic distance, this energy is of the order of several electron volts.

The important conclusion of this calculation is that the electrons in a metal at the absolute zero of temperature are not at rest, but have energies between zero and some value $W_{\max.}$ of this order, and hence large compared with kT. This band of energy levels is known as the Fermi distribution of levels.

The calculation can be extended to three dimensions; the formula for $W_{\max.}$ is then

$$W_{\max.} = \frac{\pi^2\hbar^2}{2m}\left(\frac{3N}{\pi}\right)^{\frac{2}{3}},$$

where N is the number of electrons per unit volume in the metal.

It will be seen that, since $W_{max.} \gg kT$, raising the temperature will excite only a fraction of the total number of electrons of order $kT/W_{max.}$, and that these will each acquire energy of the order kT. Thus the total thermal energy of the N electrons in a metal is a numerical constant multiplied by

$$N(kT)^2/W_{max.}$$

The specific heat per electron is thus a multiple of

$$k^2 T/W_{max.}$$

and thus linear in the temperature.

The electronic specific heat is important at low temperatures, since that due to lattice vibrations varies as T^3. It has been observed at the temperature of liquid helium for a number of metals.†

An exact discussion of the specific heat involves a treatment of Fermi-Dirac statistics, the statistics obeyed by particles such as electrons for which wave-functions must be anti-symmetrical. For this the reader is referred to text-books on statistical mechanics.

The most direct experimental proof of the existence of a broad band of energy levels is provided by the form of the X-ray emission *bands* which result when an electron makes a transition from the conduction band to an X-ray level (Fig. 27a). The width of the band gives directly the width of the band of occupied levels.‡

We show in Fig. 27b the usual energy level scheme for a metal; the band of occupied levels of width $W_{max.}$ (the Fermi distribution) and the work function ϕ representing the minimum energy required to remove an electron from the metal. This can be determined either photoelectrically or from thermionic emission.

† Cf. N. F. Mott and H. Jones, *Theory of the Properties of Metals and Alloys*, 1936, pp. 182, 193.

‡ For a review of this subject from the experimental point of view see, for example, H. W. B. Skinner, *Rep. Prog. Phys.* v, 271, 1938.

Fig. 27c shows the energies in the presence of a field F pulling electrons away from the surface. OA represents the potential energy $-eFx$ of an electron in the field. If the field is strong enough, electrons can be pulled out of the metal *through*

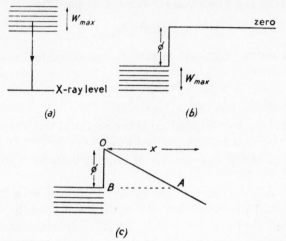

(a)   (b)

(c)

Fig. 27. Energy levels of electrons in metals. (a) X-ray emission; (b) Surface of a metal; (c) Strong field emission.

the potential barrier BOA, by 'tunnel effect'. The calculation of the chance of penetration through the barrier can be made by the method of Chapter II, § 6; the chance, that an electron with energy $W_{max.}$ incident on the barrier will penetrate it, is $e^{-\beta}$, where

$$\beta = 2 \int_0^{\phi/eF} \sqrt{\left\{\frac{2m}{\hbar^2}(\phi - eFx)\right\}}\, dx$$

$$= \frac{4}{3}\left(\frac{2m}{\hbar^2}\right)^{\frac{1}{2}} \frac{\phi^{\frac{3}{2}}}{eF}.$$

The current emitted per unit time is obtained by multiplying this by an approximately constant factor. One thus obtains the result that the current depends on the field F through a formula of the type

$$\text{current} = \text{const.}\, e^{-F_0/F},$$

where F_0 is a constant.

CHAPTER VI

TRANSITION PROBABILITIES

1. GENERAL PRINCIPLES

The problem to be treated in this chapter is the following. An atom is originally in a given stationary state, for example, the ground state. It is then perturbed by a passing charged particle, by a light wave or in some other way. After the particle has passed, or after the light wave has irradiated the atom for a certain time, the atom *may* have made a transition to one of the other stationary states. Wave mechanics enables us to calculate the probability P that this has occurred.

The principles by which such a calculation can be made are as follows. We denote the coordinates of the electron or electrons in the atom by q, and write the Schrödinger equation in the form

$$(H - W_n)\psi_n(q) = 0,$$

so that the quantities W_n are the energies that the atom can have, and $\psi_n(q)$ is the wave function describing the atom when the energy is W_n. Introducing the time factor, the complete wave function is

$$\Psi_n(q;t) = \psi_n(q)\, e^{-iW_n t/\hbar}.$$

This satisfies the time-dependent wave equation

$$-\frac{\hbar}{i}\frac{\partial \Psi}{\partial t} = H\Psi,$$

of which the general solution is

$$\sum_n A_n \psi_n(q)\, e^{-iW_n t/\hbar}, \tag{1}$$

where the A_n are arbitrary constants. Since, however, in the problem to be considered the atom is initially in the ground state, we must take for the initial form of the wave function

$$\Psi_0(q;t) = \psi_0(q)\, e^{-iW_0 t/\hbar}. \tag{2}$$

We then introduce the potential energy of the electrons in the perturbing field, for example, that of the passing particle or light wave. Explicit forms are given in (7) and (22). The potential energy will vary with the time; we write it $V(q; t)$. The wave equation for the electrons in the atom then becomes

$$-\frac{\hbar}{i}\frac{\partial\Psi}{\partial t} = H\Psi + V(q;t)\Psi. \qquad (3)$$

This equation, being linear in the time, serves to define Ψ at all subsequent times, since we are given the initial form Ψ_0 of Ψ by (2). We may expand this wave function in the form (1)— as indeed we may expand any arbitrary function of q in a series of the characteristic functions $\psi_n(q)$ of the unperturbed atom; but the coefficients A_n will now be functions of the time. We therefore write the expansion of the wave function Ψ at time t in the form

$$\Psi = \sum_n A_n(t)\,\psi_n(q)\,e^{-iW_n t/\hbar}. \qquad (4)$$

The coefficients $A_n(t)$ may be calculated; this will be done in the next paragraph. First, however, we are concerned with their *interpretation*. Their interpretation—a new physical assumption mentioned already in Chapter III of this book— is that $|A_n(t)|^2$ is the probability at time t that, under the influence of the perturbation, the atom is in the stationary state n. A wave function such as (1) or (4), in which all or several normal modes are excited, describes a state of affairs in which we do not know in which state the atom is; we are only given the probability that it is in one state or another.

Exercise

Prove that $\sum_n |A_n(t)|^2$ does not vary with the time.

We shall now show how to calculate the coefficients. If we substitute (4) into (3), we obtain

$$-\frac{\hbar}{i}\sum_n \frac{dA_n(t)}{dt}\,\psi_n(q)\,e^{-iW_n t/\hbar} = V\Psi,$$

and hence, multiplying by $\psi_n^* e^{iW_n t/\hbar}$, and integrating over all space, we find

$$\frac{dA_n(t)}{dt} = -\frac{i}{\hbar} e^{iW_n t/\hbar} \int \psi_n^* V \Psi dq. \tag{5}$$

Everything in the right-hand side is known except, of course, Ψ. An approximate solution may be obtained if we assume that the effect of the perturbing field is so small in the time considered that all the coefficients A_n remain small compared with A_0, so that the probability of excitation is small. In this case we may replace Ψ in the right hand of (5) by its original form Ψ_0; we have therefore

$$\frac{dA_n(t)}{dt} = -\frac{i}{\hbar} e^{i\omega_{n0} t} \int \psi_n^* V \psi_0 dq, \tag{6}$$

where
$$\omega_{n0} = (W_n - W_0)/\hbar.$$

The value of A_n at time t can be obtained by integrating from 0 to t; $|A_n|^2$ then gives the probability that at time t the atom is in the state n.

The approximation used here should give accurate results for the excitation of an atom by a light wave; one can always take the time short enough for all the coefficients A_n to be small compared with A_0, unless the intensity of radiation is so great that the probability of excitation is no longer proportional to E^2. For the case of a passing particle, the approximation will no longer be good for close collisions.

2. EXCITATION OF AN ATOM BY A PASSING PARTICLE

As an example we shall work out in this section the probability of excitation of a hydrogen atom by a passing particle, such as a proton. A proper treatment† describes the incident particle as well as the electron by a wave function; but the method given here, in which the proton is treated as a moving centre of force, does in fact give the correct answer if the mass of the particle is great compared with that of the electron.

† Cf. Mott and Massey.

In Fig. 28 the atom is situated at the point O, the particle moves along the line AB and passes the point C at the time $t = 0$. At time t, then, its coordinates are $(vt, p, 0)$. Here v is the velocity of the particle, and p the 'impact parameter' or

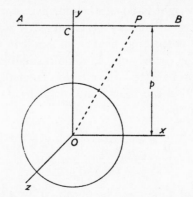

Fig. 28. Showing the excitation of a hydrogen atom at O by a passing proton moving along the line AB.

distance between the path of the particle and the nucleus of the atom. The perturbing energy V is thus the potential energy of an electron at the point (x, y, z) in the field of the proton, so that

$$V(x, y, z; t) = \frac{-e^2}{\sqrt{\{(vt - x)^2 + (p - y)^2 + z^2\}}}. \tag{7}$$

For the purpose of this example we shall consider only distant collisions, so that p is large in comparison with x, y, z. We can then approximate for V. It may be written

$$V = \frac{-e^2}{\sqrt{(R^2 + r^2 - 2Rr\cos\theta)}},$$

where R is the distance OP, and θ the angle between OP and the vector $\mathbf{r} = (x, y, z)$ giving the position of the electron. This gives, for large R,

$$V \sim -\frac{e^2}{R}\left\{1 + \frac{r\cos\theta}{R}\right\}.$$

Now $r \cos \theta$ is the projection of the vector (x, y, z) on OP; this is equal to

$$(vt \cdot x + py)/R.$$

Substituting, we find

$$V(x, y, z; t) = -\frac{e^2}{R} - \frac{e^2(vt \cdot x + py)}{R^3},$$

where

$$R = \{(vt)^2 + p^2\}^{\frac{1}{2}}.$$

We must now use this form to calculate the transition probabilities from (6). Owing to the orthogonal properties (Chap. IV, § 8) of the functions ψ_n, the first term in V gives no contribution. We see at once, integrating (6) from $t = -\infty$ to $t = +\infty$, the whole time of the collision, that

$$A_n = \frac{e^2 i}{\hbar} \int_{-\infty}^{\infty} \left[\frac{vt x_{n0} + py_{n0}}{\{(vt)^2 + p^2\}^{\frac{3}{2}}} \right] e^{i\omega_{n0} t} dt, \tag{8}$$

where

$$x_{n0} = \int \psi_n^* x \psi_0 d\tau, \quad y_{n0} = \int \psi_n^* y \psi_0 d\tau.$$

The 'matrix elements' x_{n0}, etc. have already been introduced (Chap. IV, § 10), and values given for the hydrogen atom. We have thus only to discuss the integration over t.

If the impact parameter p is large compared with v/ω_{n0}, it will be seen that the integrand oscillates a large number of times in the range of t, of order p/v, in which it is not small. Under these conditions the integral is small; it may be shown to behave as $\exp(-p\omega_{n0}/v)$. Thus as a rough approximation we may assume that all collisions are ineffective for which p is greater than this value v/ω_{n0}.

This gives an interesting formula for the distance at which collisions cease to be effective in ionising or exciting an atom. If λ $(= 2\pi c/\omega_{n0})$ is the wavelength of the radiation (light) necessary to excite or ionise the atom, the critical distance is

$$p \sim \frac{\lambda}{2\pi} \frac{v}{c},$$

so long as $v/c \ll 1$, so that no relativistic correction need be applied.

If, on the other hand, p is small compared with v/ω_{n0}, the term $\exp(i\omega_{n0}t)$ can be neglected; the integral (8) then reduces to[†]

$$A_n \sim \frac{ie^2}{\hbar} \int_{-\infty}^{\infty} \frac{py_{n0}\,dt}{\{(vt)^2+p^2\}^{\frac{3}{2}}} = \frac{ie^2}{\hbar} \frac{2y_{n0}}{pv}.$$

Thus the chance P that the atom is excited into the state n by the passing particle is given by

$$P = \frac{4e^4|y_{n0}|^2}{\hbar^2 p^2 v^2} \quad p < \frac{v}{\omega_{n0}}. \tag{9}$$

If N particles cross unit area per unit time, the chance per unit time that the atom is excited is

$$\frac{4Ne^4|y_{n0}|^2}{\hbar^2 v^2} \int \frac{2\pi p\,dp}{p^2}.$$

The upper limit of this integral may be taken to be v/ω_{n0}, and the lower a quantity of the order of the radius a of the atom, below which the approximations do not apply; thus for the mean chance per unit time that the atom is excited we obtain finally

$$\frac{8\pi Ne^4|y_{n0}|^2}{\hbar^2 v^2} \ln\left(\frac{v}{a\omega_{n0}}\right).$$

The loss of energy per unit length of path, dW/dx, for a particle going through a gas containing N_0 atoms per cm.[3] is thus

$$\frac{dW}{dx} = \frac{8\pi N_0 e^4}{\hbar^2 v^3} \sum_n \hbar\omega_{n0} |y_{n0}|^2 \ln\left(\frac{v}{a\omega_{n0}}\right), \tag{10}$$

where the summation is over all states into which a transition can occur.

These formulae express the rate of loss of energy in terms of the quantity y_{n0}, the 'dipole moment' of the transition 0 to n.

It is interesting to make an estimate of the rate of loss of energy without using quantum mechanics at all. At the moment t, the particle P exerts a force on an electron at O (Fig. 28) of which the component in the direction CO is

$$e^2 p/\{(vt)^2+p^2\}^{\frac{3}{2}}.$$

[†] Put $vt/p = \tan\theta$.

The momentum transferred to a *free* electron at O during the collision would thus be

$$\int_{-\infty}^{\infty} \frac{e^2 p}{\{(vt)^2 + p^2\}^{\frac{3}{2}}} \, dt = \frac{2e^2}{pv},$$

so that the electron if initially at rest would acquire energy

$$2e^4/mp^2v^2.$$

The corresponding quantum mechanical formula for a *bound* electron is, from (9),

$$(4e^4/p^2v^2\hbar^2) \, \Sigma \hbar \omega_{n0} \, | \, y_{n0} \, |^2,$$

where the summation is over all states n to which the electron can be excited without violating the condition $\omega_{n0} < v/p$. Since we should expect that for near collisions the absorption of energy would be about the same as for a free electron, this suggests that the sum

$$\Sigma 2m\omega_{n0} \, | \, y_{n0} \, |^2/\hbar$$

should add up to unity. This is in fact the case; the quantity f_{n0} defined by

$$f_{n0} = \frac{2m\omega_{n0}}{\hbar} \, | \, y_{n0} \, |^2 = \frac{8\pi^2 m\nu_{n0}}{h} \, | \, y_{n0} \, |^2,$$

is known as the 'oscillator strength' of the transition 0 to n, and, as already stated in Chapter IV, § 10,

$$\Sigma f_{n0} = 1.$$

Certain interesting results follow from (10). As the velocity v of an ionising particle passing through matter decreases, it will be seen that the rate of loss of energy increases, until v/ω_{n0} becomes of the same order as a. It will be seen that this is the case when the velocity of the particle becomes comparable with that of the electron in the atom. When v is very small, although the atom may be very much perturbed by the passing particle during the collision, when the particle has gone away the wave function returns to its original form. The collision is then what we call adiabatic.[†]

3. TRANSITIONS TO UNQUANTISED STATES

A perturbing field, for example, the field of a passing charged particle or light wave, may ionise as well as excite an atom. Thus we have to consider transitions also to unquantised states. In order to use the formulae of the last section, the simplest procedure is to quantise the unquantised states by imagining the atom as enclosed in a large box of side L, say. For instance, suppose that the final state of the electron is represented by the wave function of a free particle[‡]

$$\psi_{\mathbf{k}}(\mathbf{r}) = e^{i(\mathbf{kr})}. \tag{11}$$

If we introduce our box of side L, together with the boundary condition that $\psi_{\mathbf{k}}$ shall be periodic with period L, then $\mathbf{k}$ has the values
$$\mathbf{k} = 2\pi(n_1, n_2, n_3)/L,$$

where n_1, n_2, n_3 are integers.

Therefore the number of states for which $\mathbf{k}$ lies between the limits k_1 and $k_1 + dk_1$, k_2 and $k_2 + dk_2$, k_3 and $k_3 + dk_3$ is

$$L^3 dk_1 dk_2 dk_3/(2\pi)^3. \tag{12}$$

Since the normalised wave function for the free electron is

$$L^{-\frac{3}{2}} e^{i(\mathbf{kr})},$$

it follows that the chance that after time t the electron will be found in one of the states within the limits described by (12) is

$$\frac{dk_1 dk_2 dk_3}{(2\pi)^3} \left| \int_0^t e^{i w_{n0} t} \frac{1}{\hbar} \int e^{-i(\mathbf{kr})} V \psi_0 d\tau dt \right|^2. \tag{13}$$

[†] For further references on the treatment of collisions by the method of impact parameters, see the following: Niels Bohr, 'The Penetration of Atomic Particles through Matter', *K. Danske Vidensk. Selsk.* XVIII, no. 8, 1, 1948; Mott and Massey; E. J. Williams, *Rev. Mod. Phys.* XVII, 217, 1945.

[‡] The use of the free electron wave function (11) represents, of course, an approximation. One ought to replace it by the wave function of an electron with positive energy moving in the field of the atom. For details, cf. Mott and Massey, chap. XIV, § 2·1.

Alternatively one may require the probability that the electron is ejected in a direction lying in a solid angle $d\Omega$ and with energy lying between W and $W + dW$. It will easily be verified, since $k = 2\pi\sqrt{(2mW)}/h$, that the number $\rho(W)dW$ of states in this range is given by

$$\rho(W)dW = L^3 k^2 dk d\Omega/(2\pi)^3$$
$$= L^3 m^2 v d\Omega dW/h^3, \tag{14}$$

where $v(= \sqrt{(2W/m)})$ is written for the velocity of the ejected electron.

Thus the chance per unit time that the electron is ejected with energy between W and $W + dW$, and its direction of motion in the solid angle $d\Omega$, is

$$\frac{m^2 v\, d\Omega\, dW}{h^3} \left| \int_0^t e^{i\nu_{n0} t} \frac{1}{\hbar} \int e^{-i(\mathbf{kr})} V \psi_0 d\tau\, dt \right|^2.$$

4. TRANSITIONS DUE TO A FORCE ON AN ELECTRON WHICH IS PERIODIC IN THE TIME.

This case is of great importance because it includes the action of a light wave on an atom. If ν is the frequency with which the force changes, we shall show that transitions only occur with finite probability when the energy of the atom changes by $h\nu$. We have thus in this section to deduce from wave mechanics the well-known expressions of Bohr and Einstein for the absorption of energy by radiation, on which the quantum theory was first built up.

It is convenient to write the perturbing term $V(q; t)$ in the wave equation (3) in the form

$$V = V_0 e^{-i\omega t} + \text{complex conjugate}$$

where $V_0(q)$ is some function of the coordinates q of the electrons and $\omega = 2\pi\nu$. Then (6) shows that the chance P_n that, after time t, the atom is in the state n, is given by

$$P_n = |A_n(t)|^2,$$

9

where

$$-\frac{\hbar}{i}\frac{dA_n}{dt} = (n \mid V_0 \mid 0)\, e^{i(\omega_{n0}-\omega)t} + (n \mid V_0^* \mid 0)\, e^{i(\omega_{n0}+\omega)t} \qquad (15)$$

and where $\qquad (n \mid V_0 \mid 0) = \int \psi_n^* V_0 \psi_0 d\tau.$

This formula shows clearly that, unless $\omega = \pm \omega_{n0}$, A_n does not increase with the time. The sudden application of the field at time $t = 0$ will give a certain small probability for transitions into any excited state;† but thereafter the probability will not increase unless the condition stated above is satisfied. But this condition may be written

$$h\nu = \mid W_n - W_0 \mid. \qquad (16)$$

We have shown therefore that transitions will only take place effectively if the Bohr frequency condition is satisfied.

Suppose that $W_n > W_0$; that is to say that we are dealing with a transition from a lower to a higher state. Then ω_{n0} is positive; we have only to consider the first term in the right-hand side of (15); thus integrating with respect to the time, and putting in the condition that A_n should vanish when $t = 0$, we have

$$A_n(t) = \frac{1}{i\hbar}(n \mid V_0 \mid 0)\frac{\exp\{i(\omega_{n0}-\omega)t\}-1}{i(\omega_{n0}-\omega)}.$$

Thus, squaring,

$$P_n(t) = \frac{1}{\hbar^2}\mid(n \mid V_0 \mid 0)\mid^2 \frac{2[1-\cos\{(\omega_{n0}-\omega)t\}]}{(\omega_{n0}-\omega)^2}. \qquad (17)$$

If we now set ω equal to ω_{n0}, the probability $P_n(t)$ increases with the *square* of the time, and not linearly with the time as we should expect on physical grounds. This apparent contradiction, however, can be resolved as follows.

If the final state n is quantised, the concept of radiation of *exactly* the frequency given by (16) acting on the atom is incorrect. The absorption lines have a certain width; the problem of physical importance is to calculate the chance of excitation

† The fact that the exciting radiation begins at a certain time means that its Fourier analysis must contain frequencies other than ν.

of the atom by radiation covering a band of frequencies wide compared with this. This is done in the next section, and it is shown that the probability increases linearly with the time. If the final state is unquantised (i.e. if the calculation refers to the *ionisation* of an atom by a light wave) we may use the fictitious quantisation of the last section. The atom is enclosed in a 'box' of side L, and the matrix element becomes

$$(n \mid V_0 \mid 0) = \frac{1}{L^{\frac{3}{2}}} \int e^{-i(\mathbf{kr})} V_0(q) \, \psi_0(q) \, d\tau.$$

The number of states $\rho(W) \, dW$ with energies between W and $W + dW$ is given by (14). The chance that after a time t an electron is ejected from the atom into the solid angle $d\Omega$ with any energy whatever is then given by the integral

$$\int P_n(t) \, \rho(W_n) \, dW_n,$$

the integral being over all energies W_n. The quantity $P_n(t)$ defined by (17) has, however, a strong maximum when $\omega_{n0} = \omega$, ωt being large; and thus for large t the integral becomes

$$\left[\frac{1}{\hbar} \mid (n \mid V_0 \mid 0) \mid^2 \rho(W) \right]_{W = W_0 + h\nu} \times \int \frac{2[1 - \cos\{(\omega_{n0} - \omega) t\}]}{(\omega_{n0} - \omega)^2} \, d\omega_{n0}.$$

We have here put $dW_n = \hbar \, d\omega_{n0}$ and taken outside the integral sign all terms except the one which gives rise to the sharp maximum.

To evaluate the integral we put $(\omega_{n0} - \omega) t = x$; we obtain

$$t \int_{-\infty}^{\infty} \frac{2(1 - \cos x)}{x^2} \, dx.$$

The integral is equal to 2π; thus finally the chance that an electron is ejected with velocity $v \ (= \hbar k/m)$ lying in the solid angle $d\Omega$ is, per unit time

$$\frac{2\pi}{\hbar} \{ \rho(W) \mid (n \mid V_0 \mid 0) \mid^2 \}_{W = W_0 + h\nu},$$

or

$$d\Omega \frac{4\pi^2 m^2 v}{h^4} \left| \int e^{-i(\mathbf{kr})} V_0 \psi_0 \, d\tau \right|^2, \tag{18}$$

5. EMISSION AND ABSORPTION OF RADIATION

Planck in his theory of black-body radiation was the first to introduce the hypothesis that the energy of a light wave of frequency ν is quantised, being a multiple of $h\nu$. Einstein explained the photoelectric effect by applying this idea to the interaction between radiation and matter; light can only give up its energy to matter in quanta of amount $h\nu$. Niels Bohr in his theory of the hydrogen spectrum introduced the complementary idea, that an atom in making a transition from one quantised state to another would radiate quanta of energy of frequency determined by (16).

Einstein was the first to give a correct quantitative description of the intensities of radiative processes. Corresponding to the transitions between any two states n and m of an atomic system, he introduced three probability coefficients, A_{nm}, B_{nm}, and B_{mn}. The coefficient A_{nm} is defined as follows. $A_{nm}dt$ is the probability per time dt that an atom initially in the state n will make a transition to the state m, in the absence of all perturbation from outside. The coefficient A_{nm} is supposed to be independent of t, of the past history of the atom and of the process by which it has been brought to the state n.

The coefficients B_{nm} and B_{mn} are defined as follows. Suppose the atom is in the presence of radiation, unpolarised and incident in all directions with equal intensity, and such that the energy per unit volume with frequency between ν and $\nu+d\nu$ is $I(\nu)\,d\nu$. Let ν_{nm} be the frequency corresponding to a transition between the states n, m. Then if the atom is in the state m (the lower state) the chance per time dt that it will make a transition to the upper state n with the absorption of a quantum of radiation is $B_{mn}I(\nu_{nm})\,dt$. Also, if it is in the upper state n, the chance that in the presence of radiation it will make a transition to the ground state with the emission of a quantum is

$$\{A_{nm}+B_{nm}I(\nu_{nm})\}\,dt.$$

In attempting to calculate the A and B coefficients by means of wave mechanics we are faced with the following difficulty.

The B coefficients may be calculated by a direct application of the analysis already given, the light wave being treated as an alternating electromagnetic field; so also may the probability of the photoelectric effect, the ejection of an electron from an atom. The A coefficient, however, cannot be calculated in this way; at first sight there seems to be no perturbation which will cause a spontaneous transition.

A more elaborate theory than will be developed in this book is necessary to account for the A coefficient. Such a theory was first given by Dirac.† In this theory the radiation field is treated as a quantised vibrating system which interacts with the atoms, and, even when no radiation quanta are present, the interaction enables an excited atom to jump to a lower state and create one. For further details of this theory, the reader is referred either to the original papers or to more advanced text-books.‡

Fortunately, the A and B coefficients are connected by equations which depend on thermodynamics. These are

$$B = B_{nm} = B_{mn}, \quad A = \frac{8\pi h\nu^3}{c^3}B. \qquad (19)$$

They may be proved as follows. Suppose that a number of atoms of the type concerned are in thermal equilibrium in an enclosed space at temperature T, together with the black-body radiation characteristic of that temperature. Then in a given interval of time as many atoms must make the transition upwards as downwards. Thus if $I(\nu)$ refers to the intensity of black-body radiation, and N_n, N_m are the numbers of atoms in the states n, m,

$$N_n\{A_{nm} + B_{nm}I(\nu_{nm})\} = N_m B_{mn}I(\nu_{nm}).$$

But by Boltzmann's law

$$N_n/N_m = \exp{(-h\nu_{nm}/kT)}.$$

† P. A. M. Dirac, *Proc. Roy. Soc.* A, cxiv, 243, 1927.
‡ E.g., W. Heitler, *Quantum Theory of Radiation*, 2nd ed., 1944.

Now if we make the temperature tend to infinity, N_n and N_m become equal and $I(\nu)$ becomes large; it follows that the two B coefficients are equal. We may therefore write, dropping the suffixes n, m,

$$I(\nu) = \frac{A}{B} \frac{1}{e^{h\nu/kT} - 1}. \tag{20}$$

We know, however, from the application of statistical mechanics to the radiation itself that the density of black-body radiation is quite independent of the type of atom present, and is in fact given by

$$I(\nu) = \frac{8\pi h\nu^3}{c^3} \frac{1}{e^{h\nu/kT} - 1} \tag{21}$$

Comparing (20) and (21), (19) follows.

We shall therefore, in the next section, limit ourselves to the calculation of B, the coefficient of absorption or forced emission.

6. CALCULATION OF THE B COEFFICIENT

We make use of the method of § 4. The simplest approach will be to treat the light as an oscillating electric field $E_0 \cos \omega t$ and neglect the effect of the magnetic field on the electron. A more rigorous approach which gives approximately the same result will be mentioned at the end of this section.

We take the electric vector E along the z-axis; then the potential energy of the electron is

$$V(x, y, z; t) = eEz = eE_0 z \cos \omega t. \tag{22}$$

This, as in § 4, can be written

$$V = V_0 e^{-i\omega t} + V_0^* e^{i\omega t},$$

where

$$V_0 = V_0^* = \tfrac{1}{2} eE_0 z.$$

Thus (15) can be integrated, subject to the condition that $A_n(t) = 0$ when $t = 0$, to give

$$-\frac{\hbar}{i} A_n(t) = \tfrac{1}{2} eE_0 z_{n0} \left[\frac{\exp\{i(\omega_{n0} - \omega) t\} - 1}{i(\omega_{n0} - \omega)} + \frac{\exp\{i(\omega_{n0} + \omega) t\} - 1}{i(\omega_{n0} + \omega)} \right], \tag{23}$$

where z_{n0} denotes the matrix element (already introduced on pp. 75, 125)

$$z_{n0} = \int \psi_n^* z \psi_0 d\tau.$$

We now consider the two terms within the square brackets. If ω is not equal or very nearly equal to $\pm \omega_{n0}$, the function oscillates with t; it does not increase. If $\omega - \omega_{n0}$ vanishes, however, it increases with t. Thus, if we are dealing with absorption, so that the initial state is below the final and $\omega_{n0}(= \overline{W_n - W_0}/\hbar)$ is positive, we need consider only the former term and values of ω near to ω_{n0}. In the case of stimulated emission the initial state of energy W_0 is above the final state, ω_{n0} is negative and only the second term need be considered.

In the former case, that of absorption, we retain the first term in (23). Taking the square, we find

$$| A_n(t) |^2 = \left(\frac{e E_0 z_{n0}}{2\hbar} \right)^2 \frac{2[1 - \cos\{(\omega_{n0} - \omega)t\}]}{(\omega_{n0} - \omega)^2}.$$

This function, giving the chance that after time t the atom is in the state n, has as we expect a strong maximum for $\omega_{n0} = \omega$. To obtain an expression for the absorption coefficient, we suppose that the atom is irradiated with light such that the energy density between the frequencies ν, $\nu + d\nu$ is $I(\nu) d\nu$. The energy density in the light wave with electric vector $E_0 \cos \omega t$ is the mean value of $(E^2 + H^2)/8\pi$, or of $E^2/4\pi$ and thus $E_0^2/8\pi$. Replacing E_0^2 by $8\pi I(\nu) d\nu$ in the expression above for $| A_n(t) |^2$, we see that the chance that after time t the atom is excited into the state n is

$$P = 2\pi \left(\frac{e z_{n0}}{\hbar} \right)^2 \int I(\nu) \frac{2[1 - \cos\{(\omega_{n0} - \omega)t\}]}{(\omega_{n0} - \omega)^2} d\nu.$$

Owing to the strong maximum at $\omega = \omega_{n0}$, we may take $I(\nu)$ outside the integral sign. The substitution of p. 131 enables the integral to be evaluated. Writing

$$d\nu = d\omega/2\pi, \quad (\omega_{n0} - \omega)t = x,$$

we obtain
$$P = \left(\frac{e z_{n0}}{\hbar} \right)^2 I(\nu_{n0}) t \int_{-\infty}^{\infty} \frac{2(1 - \cos x)}{x^2} dx.$$

The integral as before is equal to 2π; thus

$$P = 8\pi^3 \left(\frac{ez_{n0}}{h}\right)^2 I(\nu_{n0})\, t,$$

which increases with t as it should.

To obtain the B coefficient we must average over all directions of polarisation; we obtain

$$B = \frac{8\pi^3}{3}\frac{e^2}{h^2}\left(\,|\,x_{n0}\,|^2 + |\,y_{n0}\,|^2 + |\,z_{n0}\,|^2\right). \tag{24}$$

The A coefficient may be obtained from (19), and is

$$A = \frac{64\pi^4}{3}\frac{e^2\nu^3}{hc^3}\left(\,|\,x_{n0}\,|^2 + |\,y_{n0}\,|^2 + |\,z_{n0}\,|^2\right). \tag{25}$$

It is of interest to express this in terms of the oscillator strength already introduced on pp. 76, 127; we find, since normally all but one of the three matrix elements in (25) will vanish,

$$A = \frac{8\pi^2}{3}\frac{e^2\nu^2}{mc^3}f. \tag{26}$$

If one inserts numerical values one finds

$$A = 2\cdot 2 \times 10^8 \frac{f}{\Lambda^2}\, \text{sec.}^{-1},$$

where Λ is the wavelength in μ. For a line of strong intensity f lies, say, between $0\cdot 1$ and 1. Thus the transition probability is about 10^8 sec.$^{-1}$ for a line in the visible region, but much more for lines in the X-ray region.

Finally we must add a few remarks about the approximation (22) for the field of the light wave. Here we neglect both the effect of the magnetic field, and the variation of E within the atom. The correct perturbing term is actually, for a light wave moving along the z-axis with electric vector along the x-axis

$$\frac{e\hbar E_0}{4\pi i m\nu}\left\{e^{i\omega(z/c-t)}\frac{\partial}{\partial x} + \text{complex conjugate}\right\}.$$

This gives the same results if the radius of the atom is considered small in comparison with c/ω.

7. SELECTION RULES

Optical transitions from a state n to a state m are said to be forbidden when the three matrix elements x_{mn}, y_{mn}, z_{mn} all vanish. This will be the case in the following circumstances. Suppose that in the state m the number of quanta of angular momentum is l, and the component along the z-axis u; in other words, suppose that the wave function is of the form

$$f(r)P_l^u(\cos\theta)\,e^{iu\phi}.$$

Suppose also that for the state n the corresponding quantities are l', u'. Then the matrix elements x_{mn}, etc. will vanish *unless*

$$l-l' = \pm 1, \quad u-u' = 0 \text{ or } \pm 1. \tag{27}$$

To prove this we have to show that the integral

$$\iint (\alpha x + \beta y + \gamma z)\,P_l^u(\cos\theta)\,P_{l'}^{u'}(\cos\theta)\,e^{u(u-u')\phi}\sin\theta\,d\theta\,d\phi$$

vanishes unless (27) is satisfied, α, β, γ being arbitrary direction cosines. Since x, y, z may be written $r\sin\theta\cos\phi$, $r\sin\theta\sin\phi$, $r\cos\theta$, the proof is obvious for the selection rule in u. For the selection rule in l the reader is referred to text-books on spherical harmonics.

For transitions involving s and p states, the selection rule may be verified easily. For instance, s states have spherically symmetrical wave functions. The matrix element of z for the transition between two s states will thus necessarily be of the form

$$\iiint z \times (\text{a function of } r)\,dx\,dy\,dz.$$

This will vanish by symmetry, values for positive z just cancelling values for negative z. For a transition between an s and a p state, on the other hand, the matrix elements do not vanish. If the p state has a wave function of the form $xf(r)$, then y_{mn}, z_{mn} vanish, but x_{mn} does not.

Transitions for which x_{mn}, etc. vanish are not forbidden absolutely, since the formulae (24), (25) are based on the approximation that the field in the atom can be taken to be $E_0 \cos \omega t$, and thus constant throughout the atom. Actually it is of the form $E_0 \cos \{\omega(t - x/c)\}$, which differs from that assumed by a term of order $\omega a/c$, where a is the radius of the atom. Thus if the matrix elements x_{mn} vanish, the transition probability is of order $(\omega a/c)^2$ smaller than the values calculated here. Since $c/\omega = \lambda/2\pi \sim \frac{1}{2} . 10^{-5}$ cm., and $a \sim 10^{-8}$ cm., it will be seen that forbidden transitions are c. 10^5 less probable than those optically allowed.

8. PHOTO-ELECTRIC EFFECT

Formula (18) may be used to calculate the probability that an electron is ejected from an atom by a light wave of suitable frequency. As before, we suppose that the electric vector of the light wave is along the z-axis, and that the force on the electron in this direction is $E_0 e \cos \omega t$. Then (18) gives for the chance per unit time that an electron is ejected in a solid angle $d\Omega$ lying about the direction of the vector $\mathbf{k}$

$$e^2 E^2 \frac{\pi^2 m^2 v}{h^4} d\Omega \left| \int e^{-i(\mathbf{kr})} z \psi_0 d\tau \right|^2. \tag{28}$$

We shall evaluate the integral for the case when ψ_0 is the wave function of an electron in an s state, and thus spherically symmetrical. We shall take spherical polar coordinates with axis along the direction of $\mathbf{k}$, and shall denote by α the angle between this direction and the z-axis. Then

$$\int e^{-i(\mathbf{kr})} z \psi_0 d\tau = 2\pi \int_0^\pi \int_0^\infty e^{-i\,kr\cos\theta} \cos\theta \cos\alpha\, \psi_0(r) \sin\theta\, d\theta\, r^3 dr.$$

Integration over θ gives

$$4\pi \cos \alpha \int_0^\infty \left\{ \frac{\cos kr}{kr} - \frac{\sin kr}{(kr)^2} \right\} \psi_0(r)\, r^3 dr.$$

One result of importance is that no electrons are ejected in the direction perpendicular to E, since $\cos\alpha$ then vanishes. However, we shall not give details of the evaluation of the integrals, because the approximations used in (24) make the result of qualitative value only. In the first place, for low velocities of the ejected electron, it is not a good approximation to use $e^{i(\mathbf{kr})}$ for its wave function; one should use a solution of the Schrödinger equation for an electron in the field of the nucleus, normalised to behave at infinity like $e^{i(\mathbf{kr})}$. Secondly, for high velocities of the ejected electron, it is not sufficient to represent the light wave by $E_0 \cos\omega t$, as explained at the end of the last section.†

9. TRANSITIONS DUE TO A PERTURBATION WHICH DOES NOT VARY WITH THE TIME

9·1. *Auger effect*

If an electron is removed from an X-ray level of an atom, for instance the K level, two events may occur:

(a) An electron in a higher level, for instance in an L level, may make a transition to the K level, emitting a quantum of X-radiation.

(b) Alternatively an electron in an L or higher level may take up the energy, and itself be ejected out of the atom. Such a process is known as the Auger effect. The purpose of this section is to outline the calculation of the chance per unit time that it occurs.

We shall define the problem as follows. Initially we have an electron in an L level, which we shall take to be an L_2 or L_3 level, so that the wave function is of the form

$$\psi_{2p}(\mathbf{r}) = f(r)z,$$

and transitions to the K level with wave function $\psi_{1s}(\mathbf{r})$ are allowed. The second electron, which is to absorb the energy,

---

† For a more detailed discussion of the photo-electric effect, see Mott and Sneddon, p. 266; W. Heitler, p. 122; H. Bethe, *Handb. Phys.* XXIV, pt. 1, 475, 1933; M. Stobbe, *Ann. Phys.* VII, 661, 1930.

we shall take to be in a state $\psi_0(\mathbf{r})$ of which the symmetry may be defined later. We shall suppose that by taking up the energy of the first electron, it makes a transition to the free state with wave function $e^{i(\mathbf{kr})}$. The initial state of the system is that described by the wave function

$$\psi_{2p}(\mathbf{r}_1)\,\psi_0(\mathbf{r}_2),$$

and the final state by the wave function

$$\psi_{1s}(\mathbf{r}_1)\,e^{i(\mathbf{kr}_2)}.$$

If there were no interaction between the two electrons, the Auger transition would not occur. We therefore make use of the analysis of §§ 3, 4 and calculate the probability per unit time that the transition will occur, under the influence of the interaction energy

$$V(\mathbf{r}_1, \mathbf{r}_2) = \frac{e^2}{|\,\mathbf{r}_1 - \mathbf{r}_2\,|} = \frac{e^2}{\sqrt{\{(x_1 - x_2)^2 + (y_1 - y_2)^2 + (z_1 - z_2)^2\}}}.$$

This interaction energy does not contain the time; we thus have to put $\omega = 0$ in the analysis of § 4. We find for the probability per unit time that the transition occurs, the electron being ejected into a solid angle $d\Omega$ about the direction $\mathbf{k}$,

$$\frac{4\pi^2 m^2 v\, d\Omega}{h^4} \left| \iint \psi_{1s}^*(\mathbf{r}_1)\, e^{-i(\mathbf{kr}_2)}\, \frac{e^2}{|\,\mathbf{r}_1 - \mathbf{r}_2\,|}\, \psi_{2p}(\mathbf{r}_1)\, \psi_0(\mathbf{r}_2)\, d\tau_1 d\tau_2 \right|^2 \quad (29)$$

We shall suppose that the radial extension of the wave function ψ_{1s} of the K level is small compared with that of ψ_0, the initial state of the electron that is to be ejected. We may then neglect values of r_2 less than r_1 in the integral. Thus $1/|\,\mathbf{r}_1 - \mathbf{r}_2\,|$ may be expanded

$$\frac{1}{|\,\mathbf{r}_1 - \mathbf{r}_2\,|} = \frac{1}{r_2}\left[1 + \frac{r_1}{r_2}\cos\Theta + \dots\right],$$

where Θ is the angle between $\mathbf{r}_1, \mathbf{r}_2$. We have

$$\cos\Theta = \cos\theta_1 \cos\theta_2 + \sin\theta_1 \sin\theta_2 \cos(\phi_1 - \phi_2).$$

Thus carrying out the integration over $d\tau_1$, we have, making use of the orthogonal properties of ψ_{1s}, ψ_{2p},

$$\int \psi_{1s}^*(\mathbf{r}_1) \psi_{2p}(\mathbf{r}_1) \frac{1}{|\mathbf{r}_1 - \mathbf{r}_2|} d\tau_1 = \frac{M z_2}{r_2^3},$$

where M is the matrix element,

$$M = \int \psi_{1s}^* \psi_{2p} z \, d\tau.$$

We are thus left with

$$\frac{4\pi^2 m^2 v e^4 d\Omega}{h^4} M^2 \left| \int e^{-i(\mathbf{kr})} \frac{z}{r^3} \psi_0(\mathbf{r}) \, d\tau \right|^2. \tag{30}$$

No attempt will be made here to work out the integral, though the student may easily carry it out with $2s$ or $2p$ hydrogen-like functions. We shall confine ourselves to certain remarks about the orders of magnitude concerned. If we suppose that the radii of all the orbits concerned are of comparable magnitude r_0, then

$$v \sim h/m r_0, \quad M \sim r_0,$$

and the integral is of order $r_0^{-\frac{1}{2}}$. Thus the order of magnitude of the transition probability is $me^4/\hbar^3$, and is independent of r_0, and therefore does *not* depend on the atomic number Z. Now the probability of an optical transition increases rapidly with Z, according to (26) as Z^4, since ν varies as Z^2. Thus the Auger transitions are important for light elements, but not for heavy elements. For light elements they frequently cut down the intensity of emitted radiation by a very large factor.

Actually the transition probability may be much less than $me^4/\hbar^3$ ($\sim 10^{15}$ sec.$^{-1}$) owing essentially to the rapid oscillations of the term $e^{i(\mathbf{kr})}$ in the integral in (30). This will be the case particularly if the velocity of the ejected electron is great compared with its velocity in the state ψ_0.

9·2.   *Scattering of electrons by a centre of force*

In Chapter II, § 7, it was shown that, if a beam of particles, such that one crosses unit area per unit time, is incident on a centre of force, the chance per unit time that a particle is scattered through an angle θ into the solid angle $d\Omega$ is

$$\left| \frac{2\pi m}{h^2} \iiint e^{i\mu y} V(r) \, dx \, dy \, dz \right|^2 d\Omega,$$

where
$$\mu = 2k \sin \tfrac{1}{2}\theta.$$

This formula may be deduced at once from (18) by taking for ψ_0 the wave function describing electrons moving in the direction of a vector $\mathbf{k}_0$, again one crossing unit area per unit time,

$$\psi_0 = v^{-\frac{1}{2}} e^{i(\mathbf{k}_0 \mathbf{r})},$$

and treating $V(r)$ as a perturbing force which causes them to be scattered. Then (18) gives for the number scattered per unit time into the solid angle $d\Omega$

$$\left| \frac{2\pi m}{h^2} \int e^{i(\overline{\mathbf{k}_0 - \mathbf{k}} \cdot \mathbf{r})} V(r) \, d\tau \right|^2 d\Omega.$$

To evaluate the integral we take the y-axis along the vector $\mathbf{k}_0 - \mathbf{k}$. If θ is the angle between the vectors $\mathbf{k}, \mathbf{k}_0$, then

$$|\mathbf{k}_0 - \mathbf{k}| = 2k \sin \tfrac{1}{2}\theta,$$

so that
$$(\mathbf{k}_0 - \mathbf{k}) \cdot \mathbf{r} = 2ky \sin \tfrac{1}{2}\theta.$$

The required result follows.

CHAPTER VII

RELATIVISTIC AND NUCLEAR DEVELOPMENTS

1. DIRAC'S RELATIVISTIC WAVE EQUATION

No full account of Dirac's relativistic theory will be attempted here. In his original paper† on the subject Dirac showed that, in order to write down a wave equation linear in the time (it must be linear for the reasons given in Chap. III, § 2), and satisfying the principle of relativity, it was necessary to ascribe to the electron a fourth degree of freedom, or spin; he showed that, if this were done, the properties previously ascribed to the spin could be deduced without further assumptions from the equation. In any relativistic treatment x, y, z must be treated on the same footing as it; thus the equation is linear in $\partial/\partial x, \partial/\partial y, \partial/\partial z$ as well as $\partial/\partial t$. The state of the electron is described in Dirac's theory by *four* wave functions $\psi_1, \psi_2, \psi_3, \psi_4$; and the quantity $|\psi|^2$ of the non-relativistic theory is replaced by

$$\psi_1\psi_1^* + \psi_2\psi_2^* + \psi_3\psi_3^* + \psi_4\psi_4^*.$$

The relativistic theory should be used in all cases where the velocity of the electron approaches that of light, and also, even when this is not so, to calculate the separation of spin doublets. Some of the more important applications are the following:

(*a*) Calculation of separation of the ns and np states of hydrogen (they coincide only in the non-relativistic approximation) and the spin doublet separation of the np state.‡

(*b*) Calculation§ of the separation between L_I and L_{II} levels in the X-ray spectrum. These levels have quantum numbers

† P. A. M. Dirac, *Proc. Roy. Soc.* A, cxvii, 610, 1928. Cf. also *Quantum Mechanics.*

‡ Dirac, *Proc. Roy. Soc.* A, cxvii, 610, 1928, and *Quantum Mechanics,* 1947, p. 268; H. Bethe, *Handb. Phys.* xxiv, pt. 1, 311, 1933.

§ Cf. H. Bethe, *Handb. Phys.* xxiv, pt. 1, 322, 1933.

$2s, 2p$. The separation is here partly due to the relativistic correction, comparatively large for heavy nuclei since the mean velocity of the electron increases with Z, and partly to the fact that the field no longer has the Coulomb form owing to the screening effect of the other electrons.

(c) Calculations of the photoelectric effect and Compton effect for high energies of the electrons.†

(d) Calculations of the scattering of fast electrons by atomic nuclei, and of the polarisation of the electron beam produced by such scattering. In the scattered beam it may be shown that the spins are no longer oriented at random.‡

(e) Scattering of fast electrons by electrons.§

2. THE POSITRON OR POSITIVE ELECTRON

The positive electron was discovered independently by Anderson‖ and by Blackett and Occhialini.¶ It is a particle with the same mass as the electron, but carrying a positive charge. It is unstable, being produced when fast electrons or γ-rays interact with nuclei, and being capable of combining with an electron to produce one or more γ-ray quanta.

The theory given by Dirac to account for the positron is as follows. According to relativistic theory, the relation between the energy W of a particle of mass m and its momentum p is

$$W^2 = c^2(m^2c^2 + p^2). \tag{1}$$

If p is small this gives

$$W = mc^2 + \tfrac{1}{2}p^2/m + \text{terms in } (1/c^2),$$

so that the formula for the energy includes the energetic equivalent mc^2 of the mass as well as the kinetic energy $p^2/2m$.

Formula (1) gives for W

$$W = \pm c\sqrt{(m^2c^2 + p^2)}.$$

---

† A discussion of this subject is given by W. Heitler, *Quantum Theory of*
‡ Mott and Massey, chap. IV.   [*Radiation.*
§ Mott and Massey, p. 365. Original paper, C. Møller, *Z. Phys.* LXX, 786, 1931.
‖ C. D. Anderson, *Phys. Rev.* XLI, 405, 1932.
¶ P. M. S. Blackett and G. P. S. Occhialini, *Proc. Roy. Soc.* A, CXXXIX, 699, 1933.

In relativistic theory before the advent of wave mechanics it was possible to limit the physically significant values of W to the positive square root only; W could only change continuously and would not be able to pass through the forbidden values between $\pm mc^2$ to reach the negative values. In wave mechanics, on the other hand, the theory allows a particle to jump from any one allowed state to any other under the influence of a suitable perturbation. A suitable perturbation is the combined influence of radiation and of an atomic nucleus. Nothing corresponding to such transitions appeared to occur in nature; in fact it was not clear before the discovery of the positive electron exactly what significance the states of negative energy might have.

To overcome this difficulty in the theory, Dirac† made the following proposal. He suggested that the states with negative energy should all be occupied, in the same sense that the states of the Fermi distribution of a metal are all occupied (Chap. V, § 7·3). Space is thus to be filled with a uniform 'gas' of electrons of infinite density, and with a uniform distribution of positive charge to neutralise their charge. Or at any rate space was assumed to behave as if this were so.

The immediate consequence of this hypothesis is that a quantum of radiation, or a charged particle of kinetic energy greater than $2mc^2$, is capable in suitable circumstances (for example, in collision with a nucleus), of lifting an electron from the continuum into a state of positive energy; one would thus observe the creation of

(a) an electron,

(b) a vacancy in the continuum, which Oppenheimer‡ first showed would behave like a positive electron.

As is well known, the creation of such pairs is frequently observed. They can be created for instance by X- or γ-rays of frequency ν such that

$$h\nu > 2mc^2,$$

† Cf. *Quantum Mechanics*, p. 272.
‡ J. R. Oppenheimer, *Phys. Rev.* xxxv, 939, 1930.

as will be clear from Fig. 29. They cannot be formed in free space by a single quantum, since it is impossible to conserve at the same time energy and momentum; they can, however, be formed when radiation falls on matter. The calculation of the intensity of pair formation by X-rays is carried out by the method of Chapter VI, § 8; the initial and final wave functions are those for an electron, with negative and positive energy values respectively, moving in the Coulomb field of the nucleus.†

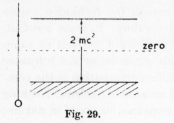

Fig. 29.

3. APPLICATIONS TO NUCLEAR PHYSICS

Applications of wave mechanics to the nucleus can only be made in certain special cases, because of our ignorance of the nature of the forces between the nucleons (protons and neutrons) of which the nucleus is believed to be made up. Problems to which wave mechanics can be applied are of three main types:

(a) Problems where, for one reason or another, only the Coulomb part of the field round a nucleus is important. In this category we may place the theory of α-decay, of the internal conversion of γ-rays and perhaps the theory of β-decay.

(b) Problems in which some assumption is made about the force between nucleons, and wave mechanics is used to calculate the consequences of the assumption. Attempts have been made along these lines to calculate the binding energy of the deuteron (one proton with one neutron), and of heavier nuclei.

(c) Problems, such as those concerned with the compound nucleus and the spacing between the energy levels of the nucleus, which do not depend on the law of force assumed.

In this book we shall discuss briefly only two of these problems, the theories of α- and of β-decay. The first is the best

† For details of the calculation, see W. Heitler, chap. IV.

known illustration of the quantum mechanical tunnel effect, the second gives an illustration of how, in a field still not completely understood, it is possible to obtain useful results from wave mechanics.

3·1.   *The theory of α-decay of radioactive elements*

The quantitative explanation by Gamow[†] and by Gurney and Condon[‡] of the emission of α-particles from radioactive nuclei was one of the earliest successes of wave mechanics. The facts to be explained are as follows. Given N atoms of a radioactive substance, $N\lambda dt$ of them will disintegrate spontaneously in a time interval dt, where λ, the decay constant, is independent of the time, and thus independent of the age of the nucleus. For different radioactive substances λ has a very wide range of values from c. 10^{-18} sec.$^{-1}$ to c. 10^6 sec.$^{-1}$; there exists a roughly linear relationship between $\log \lambda$ and the energy W with which the particle is emitted.

A nucleus of mass number $M + 4$ (i.e. containing $M + 4$ nucleons) and of atomic number $Z + 2$ will show α-decay if and only if energy is released on removing an α-particle to form a nucleus (M, Z); we shall assume this to be the case, the energy difference being W. Consider then the potential energy $V(r)$ of this α-particle at a distance r from the *product* nucleus (M, Z). At large distances

$$V(r) \sim \frac{2Ze^2}{r};$$

at small distances, comparable with the radius of the nucleus, $V(r)$ must take a form corresponding to an attractive force, because the α-particle is held within the nucleus for a long time. The form of the potential energy may be as in Fig. 30, curve (b); we have, of course, no detailed knowledge of the attractive part of the curve, and indeed the α-particle may be supposed to lose its identity within the nucleus.

† G. Gamow, *Z. Phys.* LXI, 204, 1928.
‡ R. W. Gurney and E. U. Condon, *Nature, Lond.* CXXII, 439, 1928; *Phys. Rev.* XXXIII, 127, 1929.

10*

The α-particle eventually escapes with positive energy W; within the nucleus it may be pictured as having the same energy W. Thus the α-particle within the nucleus is separated from the outside by the 'potential barrier' AOB of Fig. 30; it is possible for the α-particle to penetrate this barrier by tunnel

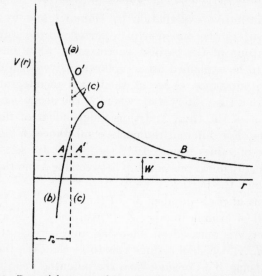

Fig. 30. Potential energy of an α-particle in the field of a nucleus.

effect. Thus the decay constant λ, the chance per unit time that the α-particle escapes, is given by

$$\lambda = \frac{v}{d_0} P,$$

where v is the velocity of the α-particle within the nucleus, d_0 is a quantity of the dimensions of its diameter and P the chance that a particle incident on the potential barrier should pass through it. The quantity v/d_0 gives an estimate of the number of times per second that the α-particle impinges on the potential barrier; no very accurate estimate need be made of it, since P varies very rapidly indeed with the various parameters.

We set $v \sim 10^9$ cm./sec., $d_0 \sim 10^{-12}$ cm., and obtain

$$\lambda = 10^{21} P \text{ sec.}^{-1}.$$

P may be calculated by the W.K.B. method (Chap. II, § 6); equation (10) of Chapter II gives

$$P = \exp\left[-2\int_A^B \left\{\frac{2m}{\hbar^2}(V-W)\right\}^{\frac{1}{2}} dr\right]. \tag{2}$$

m is here the mass of an α-particle, and the integration is from A to B in Fig. 30.

A convenient approximation for evaluation of this integral is to replace $V(r)$ by the form shown in Fig. 30, curve (c), namely

$$V(r) = \frac{2Ze^2}{r} \qquad r > r_0,$$

$$= \text{const.} \qquad r < r_0,$$

r_0 being a 'nuclear radius'. It is not suggested that $V(r)$ will actually have this form, but our complete ignorance of the true form of $V(r)$ within the nucleus, or of whether the interaction can be represented by a potential energy function at all, makes it as good an approximation as any other. Then (2) becomes

$$P = \exp\left[-2\int_{r_0}^{r_1} \left\{\frac{2m}{\hbar^2}\left(\frac{2Ze^2}{r} - W\right)\right\}^{\frac{1}{2}} dr\right],$$

where

$$r_1 = 2Ze^2/W.$$

The integral may be evaluated by setting

$$\cos^2 u = r/r_1;$$

we find

$$\ln P = -\left\{\frac{2m(2Ze^2)r_1}{\hbar^2}\right\}^{\frac{1}{2}}(2u_0 - \sin 2u_0),$$

where $\cos^2 u_0 = r_0/r_1$. Since r_0/r_1 is small, we may write

$$u_0 = \tfrac{1}{2}\pi - \sqrt{(r_0/r_1)},$$

and therefore

$$\ln P = -\frac{\pi}{\hbar}\{2mr_1(2Ze^2)\}^{\frac{1}{2}}\left(1 - \frac{4}{\pi}\sqrt{\frac{r_0}{r_1}}\right).$$

It will be seen that r_1 and Z are the most important factors in determining P, and hence the decay constant; these, obviously, define the size of the potential barrier through which the tunnel effect occurs.

Substituting for r_1, we see that a linear relation exists between $\ln P$ (and hence $\ln \lambda$) and $\sqrt{W}$ or v, the velocity with which the α-particle emerges. The relation may be written

$$\log_{10} \lambda = A - \frac{BZ}{v} + C \sqrt{r_0},$$

where

$$A = 21,$$

$$B = \frac{4\pi e^2}{\hbar \ln 10} \sim 1 \cdot 2 \times 10^9 \text{ cm./sec.}^{-1}.,$$

$$C = \frac{8(mZe^2)^{\frac{1}{2}}}{\hbar \ln 10} \sim 4 \cdot 1 \times 10^6 \text{ cm.}^{-\frac{1}{2}}.$$

The relationship between $\log \lambda$ and v, the velocity of the emitted α-particle, explains the empirical Geiger-Nuttall law, according to which the decay constants of radioactive elements increase exponentially with the energy. Also the very large variation in decay constants, from 10^5 sec.$^{-1}$ (thorium C) to $0 \cdot 5 \times 10^{-18}$ sec.$^{-1}$ (uranium) are explained by the large variation of the term BZ/v for velocities varying between about $1 \cdot 5$ and 2×10^9 cm./sec.

Detailed comparison of observed decay constants and velocities with the formula enable estimates of r_0, the nuclear radius, to be obtained. The values deduced lie between $0 \cdot 5$ and $1 \cdot 0 \times 10^{-12}$ cm.†

3·2.   *The theory of β-decay*

In Gamow's theory of α-decay explained in the last section, the decay constant λ is in principle deduced from constants already known, namely, the atomic number Z, the energy of the emitted particle, its mass m, e and $\hbar$. The unknown nuclear

† For further details of the theory of radioactive decay, see G. Gamow, *Atomic Nuclei and Nuclear Transformations*, 3rd ed., p. 174, Oxford, 1949.

radius enters only as a small correction. To develop a theory of β-decay, on the other hand, one has to make a new assumption; this is that a neutron will change into a proton with the creation of an electron and a neutrino if energy is gained thereby; or that a proton will change into a positive electron and a neutrino, again if energy is gained. And also one assumes that the probability that such a transition occurs, the electron and neutrino appearing in given quantised states with wave functions $\psi_e(x, y, z), \psi_n(x, y, z)$, is proportional to $|\psi_e \psi_n|^2$, both functions being taken at the point where the nucleon is. This theory, due originally to Fermi,† cannot account for the absolute magnitude of the decay constant; a new constant of nature has to be introduced for this; but it can account for the variation of the decay constant from element to element, and for the shape of the β-ray spectrum.

One speaks of the *creation* of an electron for the following reason. It is impossible to envisage an electron as within the nucleus before it appears outside; the radius of the nucleus is of the order 10^{-12} cm., so the wavelength of the electron would have to be of this order; it will easily be seen that such a wave-length corresponds to c. 10^8 eV, which is much greater than the energies of c. 10^6 eV with which they are actually emitted.

The neutrino is a particle with no mass or charge moving with the velocity of light. As for a light quantum, its kinetic energy W_n and momentum p_n are related by the relativistic equation, which follows from (1) when m is put equal to zero,

$$W_n = cp_n.$$

It is supposed to have a spin $\frac{1}{2}\hbar$, like an electron. It is introduced into the theory primarily in order to account for the continuous β-ray spectrum; it is supposed that a beta-active nucleus has a given amount of energy W_0 to dispose of when the neutron changes into a proton, but that this may be distributed between the electron and neutrino in any way.

† E. Fermi, Z. *Phys.* LXXXVIII, 161, 1934.

For small atomic numbers Z and high enough energies, one can describe the electron after emission by a plane wave. The neutrino can be so described in any case. Thus if the whole system is supposed shut up in a box of volume V as in Chapter VI, § 3, both ψ_e and ψ_n are of the form

$$V^{-\frac{1}{2}} e^{i(\mathbf{pr})/\hbar}.$$

Taking the nucleus at the origin, we see that ψ_e, ψ_n at that point are both independent of $\mathbf{p}$. Thus the chance per unit time for the creation of an electron and neutrino is the same for all states in which they may be found.

Now the number of states of the electron with energies between W_e and $W_e + dW_e$ is proportional to

$$p_e^2 \frac{dp_e}{dW_e} dW_e,$$

where p_e is the momentum of the electron. With each of these states is associated a number of states of the neutrino proportional to p_n^2. Thus the chance that the electron is emitted with energy between W_e and $W_e + dW_e$ is proportional to

$$p_n^2 p_e^2 \frac{dp_e}{dW_e} dW_e. \tag{3}$$

Now for p_n^2 we write $\quad p_n^2 = W_n^2/c^2$
$$= (W_0 - W_e)^2/c^2,$$

where W_0 as before is the energy available for the reaction. p_e is given in terms of W_e by the relativistic formula

$$W_e^2 = c^2(m^2 c^2 + p_e^2),$$

so that (3) becomes

$$\text{const. } \epsilon(\epsilon^2 - 1)^{\frac{1}{2}} (\epsilon_0 - \epsilon)^2 d\epsilon, \tag{4}$$

where $\quad \epsilon = W_e/mc^2 \quad$ and $\quad \epsilon_0 = W_0/mc^2.$

This formula† gives the dependence of the number of emitted β-particles in the continuous spectrum on the energy ϵ.

† For a comparison of (4) with experiment, see, for example, H. Bethe, *Elementary Nuclear Theory: A Short Course on Selected Topics*; or G. Gamow and C. L. Critchfield, *Atomic Nuclei and Nuclear Transformations*, 3rd ed., Oxford, 1949.

BIBLIOGRAPHY

1. General treatises on quantum mechanics, dealing with the applications rather than with foundations:

N. F. Mott and I. N. Sneddon, *Wave Mechanics and its Applications*, Oxford, 1948.

L. I. Schiff, *Quantum Mechanics*, New York, 1949.

Certain chapters in J. C. Slater and N. H. Frank, *Introduction to Theoretical Physics*, New York, 1933.

Certain chapters in F. K. Richtmeyer and E. H. Kennard, *Introduction to Modern Physics*, New York, 1947.

The article by H. Bethe, 'The Quantum Mechanics of the Problems of One and Two Electrons' ('Quantenmechanik der Ein- und Zwei-Elektron probleme') in *Handbuch der Physik*, vol. 24, pt. 1, 1933. This gives a very complete account of the theory of the hydrogen and helium atoms, the hydrogen molecule and their interaction with radiation.

2. General treatises on quantum mechanics, dealing with fundamentals:

P. A. M. Dirac, *Quantum Mechanics*, 3rd ed., Oxford, 1947.

The article by W. Pauli, 'General Principles of Wave Mechanics' ('Algemeine Prinzipien der Wellenmechanik') in *Handbuch der Physik*, vol. 24, pt. 1, 1933.

3. Books dealing with particular applications:

ATOMIC SPECTRA

E. V. Condon and G. H. Shortley, *The Theory of Atomic Spectra*, Cambridge, 1931.

INTERACTION BETWEEN MATTER AND RADIATION

W. Heitler, *Quantum Theory of Radiation*, 2nd ed., Oxford, 1944.

COLLISIONS BETWEEN ELECTRONS AND ATOMS

N. F. Mott and H. S. W. Massey, *Theory of Atomic Collisions*, 2nd ed., Oxford, 1949.

SOLID STATE AND MAGNETISM

F. Seitz, *Theory of Solids*, New York, 1942.

N. F. Mott and R. W. Gurney, *Electronic Processes in Ionic Crystals*, Oxford, 1940.

N. F. Mott and H. Jones, *Theory of the Properties of Metals and Alloys*, Oxford, 1936.

SOLID STATE AND MAGNETISM [*continued*]

W. Hume-Rothery, *Atomic Theory for Students of Metallurgy*, Institute of Metals, Monographs and Reports Series, 1946.

E. C. Stoner, *Magnetism and Matter*, London, 1934.

G. V. Raynor, *An Introduction to the Electron Theory of Metals*, Institute of Metals, Monographs and Reports Series, 1947.

NUCLEAR PHYSICS

G. Gamow and C. L. Critchfield, *Theory of Atomic Nucleus and Nuclear Energy Sources*, Oxford, 1949, being the third edition of Gamow's *Atomic Nuclei and Nuclear Transformations*.

H. Bethe, *Elementary Nuclear Theory. A Short Course on Selected Topics*, New York, 1947.

MOLECULAR FORCES

J. C. Slater, *Introduction to Chemical Physics*, New York, 1939.

W. G. Penney, *The Quantum Theory of Valency*, Methuen's Chemical Monographs, 1935.

J. H. van Vleck and A. Sherman, 'The Quantum Theory of Valency', article in *Reviews of Modern Physics*, VII, p. 167, 1935.

G. B. B. M. Sutherland, *Infra-red and Raman Spectra*, Methuen's Physical Monographs, 1935.

INDEX